William Thompson Sedgwick

An introduction to general biology

Second Edition

William Thompson Sedgwick

An introduction to general biology
Second Edition

ISBN/EAN: 9783337214906

Printed in Europe, USA, Canada, Australia, Japan

Cover: Foto ©berggeist007 / pixelio.de

More available books at **www.hansebooks.com**

AMERICAN SCIENCE SERIES

AN INTRODUCTION

TO

GENERAL BIOLOGY

BY

WILLIAM T. SEDGWICK, PH.D.
Professor of Biology in the Massachusetts Institute of Technology, Boston

AND

EDMUND B. WILSON, PH.D.
Professor of Zoology in Columbia College, New York

SECOND EDITION, REVISED AND ENLARGED

NEW YORK
HENRY HOLT AND COMPANY
1899

Copyright, 1886, 1895,
BY
HENRY HOLT & CO.

PREFACE TO THE FIRST EDITION.

SEVERAL years ago it was our good fortune to follow, as graduate students, a course of lectures and practical study in General Biology under the direction of Professor Martin, at Johns Hopkins University. So interesting and suggestive was the general method employed in this course which, in its main outlines, had been marked out by Huxley and Martin ten years before, that we were persuaded that beginners in biology should always be introduced to the subject in some similar way. The present work thus owes its origin to the influence of the authors of the "Elementary Biology," our deep indebtedness to whom we gratefully acknowledge.

It is still an open question whether the beginner should pursue the logical but difficult course of working upwards from the simple to the complex, or adopt the easier and more practical method of working downwards from familiar higher forms. Every teacher of the subject knows how great are the practical difficulties besetting the novice, who, provided for the first time with a compound microscope, is confronted with Yeast, Protococcus, or Amoeba; and on the other hand, how hard it is to sift out what is general and essential from the heterogeneous details of a mammal or a flowering plant. In the hope of lessening the practical difficulties of the logical method we venture to submit a course of preliminary study, which we have used for some time with our own classes, and have found practical and effective.

It has not been our ambition to prepare an exhaustive treatise. We have sought only to lead beginners in biology from familiar facts to a better knowledge of how living things are built and how they act, such as may rightly take a place in general education or may afford a basis for further studies in General Biology, Zoölogy, Botany, Physiology, or Medicine.

Believing that biology should follow the example of physics and chemistry in discussing at the outset the fundamental properties of matter and energy, we have devoted the first three chapters to an elementary account of living matter and vital energy. In the chapters which follow, these facts are applied by a fairly exhaustive study of a representative animal and plant, of considerable, though not extreme, complexity—a method which we believe affords, in a given time, a better knowledge of vital phenomena than can be acquired by more superficial study of a larger number of forms. We are satisfied that the fern and the earthworm are for this purpose the best available organisms, and that their study can be made fruitful and interesting. The last chapter comprises a brief account of the principles and outlines of classification as a guide in subsequent studies.

After this introductory study the student will be well prepared to take up the one-celled organisms, and can pass rapidly over the ground covered by such works as Huxley and Martin's "Practical Biology," Brooks's "Handbook of Invertebrate Zoölogy," Arthur, Barnes and Coulter's "Plant Dissection," or the second part of this book, which is well in hand and will probably be ready in the course of the following year.

The directions for practical study are intended as suggestions, not substitutes, for individual effort. We have striven to make the work useful as well in the class-room as in the laboratory, and to this end have introduced many illustrations. The generosity of a friend has enabled us to enlist the skill of our friend Mr. James H. Emerton, who has drawn most of the original figures from nature, under our direction. We have also been greatly aided in the preparation of the figures by Mr. William Claus of Boston.

SEPTEMBER, 1886.

PREFACE TO THE SECOND EDITION.

It was originally our intention to publish this work in two parts, the first, which appeared in 1886, being intended as an introduction, while the second was to form the main body of the work and to include the study of a series of type-forms. The pressure of other work, however, delayed the completion of the second part, and meanwhile several laboratory manuals appeared which in large measure obviated the need of it. Nevertheless the use of the introductory volume by teachers of Biology, and its sale, slowly but steadily increased. It soon appeared, however, that in some cases the work was being employed not merely as an introduction, as its authors intended, but as a complete course in itself; though the wish was often expressed that the number of types were somewhat larger. These facts, and the many obvious defects in the original volume, induced us to undertake the preparation of a second and extended edition.

With increased experience our ideas have undergone some change. We are as firmly convinced as ever that General Biology, as an introductory subject, is of the very first importance; but we are equally persuaded that it must not trespass too far upon the special provinces of Zoölogy and Botany. The present edition, therefore, differs from the original in these respects: first, while the introduction has been extended so as to include representatives of the unicellular organisms (*Amœba, Infusoria, Protococcus, Yeasts, Bacteria*), the publication of a second volume has been abandoned. It is hoped that the work as thus extended may serve a double purpose, viz., either to be used as an introduction to subsequent study in Zoölogy, Botany, or Physiology; or as a complete elementary course for general students to whom the minutiæ of these more special subjects are of less importance than the fundamental facts of vital structure and function. We believe that a sound knowledge of these facts can be conveyed by the method of study here outlined; but we must emphatically insist that neither this nor any other method will give good results unless rightly used, and that this work is not designed to be a complete text-book. Probably few teachers will find it desirable to go over the whole of the ground here laid out, and we hope that still fewer will be inclined to confine their work strictly to it. Even in a brief course the student may, after going over certain portions of this work, be made acquainted with the leading types of plants and animals; and this may be rapidly accomplished if the introductory work, however limited, has been carefully done. In extended courses we have sometimes found it desirable to postpone certain **parts of** the introductory work, returning to them at a later period.

A second modification consists in placing the study of the animal before that of the plant, which plan on the whole appears desirable, especially for students who have not been well trained in other branches of science. The main reason for this lies in the greater ease with which the physiology of the animal can be approached; for there is no doubt that beginners find the nutritive problems of the plant abstruse and difficult to grasp until a certain familiarity with vital phenomena has been attained; while most of the physiological activities of the animal can be readily illustrated by well-known operations of the human body.

The third change is the omission of the laboratory directions, these having been found unsuitable. The needs of different teachers differ so widely that it is impossible to draw up a scheme that shall answer for all. In place of the laboratory directions for students we have therefore given, in an appendix, a series of practical suggestions to teachers, leaving it to them to work out detailed directions, if desired, by the help of the standard laboratory manuals. These suggestions are the result of a good deal of experience on the part of many teachers besides ourselves, and we hope they will be found useful in procuring and preparing material (often a matter of considerable difficulty), and in deciding just what the student may reasonably be expected to do.

For the rest, the original matter has been thoroughly revised, numerous errors have been corrected, and many additions made, particularly on the physiological side.

SEPTEMBER, 1895.

TABLE OF CONTENTS.

CHAPTER I.

INTRODUCTORY.

                                                                    PAGE

Living things and lifeless things. The contrast and the likeness between living matter and lifeless matter. The journey of lifeless matter through living things. Analogy between a fountain, a flame or a whirlpool, and a living organism. Living matter is lifeless matter in a peculiar state or condition. Its characteristic properties. Biology, its scope and its subdivisions. The Biological sciences. The relations of Biology to Zoology and Botany, Morphology and Physiology. Definitions and inter-relations of the biological sciences. Psychology, Sociology. Definition of General Biology.... 1

CHAPTER II.

THE STRUCTURE OF LIVING THINGS.

Their occurrence and their size. Organisms composed of organs. Functions. Organs composed of tissues. Differentiation. Tissues composed of cells. Definitions. Unicellular organisms. Living organisms contain lifeless matter. Lifeless matter occurs in living tissues and cells. Examples. Lifeless matter increases relatively with age. Summary statement of the structure of living things. The organism as a whole—the Body—more important than any of its parts........ 9

CHAPTER III.

PROTOPLASM AND THE CELL.

Protoplasm "the physical basis of life." Historical sketch. The compound microscope and the discovery of cells in cork. The achromatic objective. The cell-theory of Schleiden and Schwann. Virchow and Max Schultze. Modern meaning of the term "cell." The discovery of protoplasm and sarcode and of their essential similarity.

Purkinje. Von Mohl. Cohn. Schultze. Appearance and structure of protoplasm. A typical cell. Its parts. Cytoplasm and the nucleus. The origin of cells. Segmentation of the egg, differentiation of the tissues, the genesis of the "body," and the physiological division of labor. Protoplasm at work. Muscular contractions. *Amœba* on its travels. "Rotation" in *Nitella* and *Anacharis*. "Circulation" of the protoplasm in hair-cells of spiderwort. Ciliary motion. The sources of protoplasmic energy. Metabolism and its phases. Vital energy does not imply a "vital force." The chemical relations of protoplasm: proteids, carbohydrates, and fats. Physical Relations: temperature, moisture, electricity, etc. The protoplasm of plants and of animals similar but not identical.................................. 20

CHAPTER IV.

THE BIOLOGY OF AN ANIMAL: THE COMMON EARTHWORM.

A representative animal. Earthworms taken as a type. Their wide distribution. The common earthworm. Its name; habitat; habits; food; castings; influence on soils; burial of objects; senses. Its differentiation: antero-posterior and dorso-ventral. Its symmetry: bilateral and serial. Plan of the earthworm's body. Organs of the body and the details of their arrangement in systems: alimentary; circulatory; excretory, respiratory; motor, nervous; sensitive; etc.. **41**

CHAPTER V.

THE BIOLOGY OF AN ANIMAL. THE COMMON EARTHWORM (Continued).

Definition of reproduction. The germ-cells. Sexual and asexual reproduction. Regeneration. The reproductive system of the earthworm. Its copulation and egg-laying. The process of fertilization, and the segmentation or cleavage of the egg. The making of the body. The gastrula. The three germ-layers: ectoblast, entoblast, mesoblast. Brief statement of the phenomena of cell-division, and of nuclear division or *karyokinesis*. The making of the organs. The fate of the germ-layers. The germ-plasm............... 72

CHAPTER VI.

THE BIOLOGY OF AN ANIMAL: THE COMMON EARTHWORM (Continued).

The microscopic anatomy or histology of the earthworm. The fundamental animal tissues and their constituent cellular elements. Epithelial, muscular, nervous, germinal, blood, and connective tissues, and their distribution in the various organs. Microscopic structure of the body-wall; of the alimentary canal; of the blood-vessels; of the dissepiments; of the nervous system, ganglia; etc.. 90

CHAPTER VII.

THE BIOLOGY OF AN ANIMAL: THE COMMON EARTHWORM (Continued.)

General Physiology. The animal and its environment. Definitions. Adaptation, structural and functional, of organism to environment. Origin of adaptations. Effect of their persistence and accumulation. Natural selection through the survival of the fittest. The need of an income of food to supply matter and energy. Nature of the income. The food and its journey through the body. Alimentation. Digestion and absorption. Circulation. Metabolism. The outgo. Interaction of the animal and the environment. Summary............. 97

CHAPTER VIII.

THE BIOLOGY OF A PLANT: THE COMMON BRAKE OR FERN.

A representative plant. Ferns taken as a type. Their wide distribution. The common brake. Its name, habitat, size, etc. General morphology of its body. Its differentiation, antero-posterior and dorso-ventral. Its bilateral symmetry. The underground stem. Origin and arrangement of the leaves. Internal structure of the rhizome and the three great tissue-systems. The elementary tissues of plants. Histology of the rhizome. Roots and branches. Embryonic tissue and the apical cell. How the rhizome grows. The frond or leaf of *Pteris* and its structure. Chlorophyll-bodies. Stomata. Veins................. 105

CHAPTER IX.

THE BIOLOGY OF A PLANT: THE COMMON BRAKE (Continued).

The various methods of reproduction in *Pteris*. Sporophore and oöphore. Alternation of generations. Sporangia. Spores. Germination of the spores. Protonema. Prothallium. The sexual organs. Antheridia. Male germ-cells. Archegonia. Female germ-cells. Fertilization. Segmentation. Differentiation of the tissues. The making of the body.. 130

CHAPTER X.

THE BIOLOGY OF A PLANT: THE COMMON BRAKE (Continued).

Physiology. The fern and its environment. Its adaptation. A definition of life. The need of an income of matter and energy. Income of *Pteris*. Its power of making foods, especially starch. The circulation of foods through the plant-body. Metabolism. Outgo. Respiration. Interaction of the fern and the environment. Special physiology of the tissue-systems and of reproduction. The question of old age. A comparison of the fern with the earthworm, and of plants in general with animals in general. The physiological importance of the chlorophylless plants..................................... 144

CHAPTER XI.

THE UNICELLULAR ORGANISMS.

The multicellular body. Its origin in continued, but incomplete, cell-division. The unicellular body. Its origin traced to complete cell-division. The multicellular body and the unicellular body as individuals. Unicellular forms physiologically "organisms." Special importance of their structural simplicity. "Organisms reduced to their lowest terms.".. 156

CHAPTER XII.

UNICELLULAR ANIMALS.

A. AMŒBA.

General Account. Habitat. Form. The "Proteus animalcule." Appearance. Pseudopodia. Locomotion. Foods. The encysted state. Structure of the unicellular body. Cytoplasm. Nucleus. Vacuoles. Reproduction by fission. Physiology. The fundamental physiological properties of protoplasm as displayed in *Amœba*. The question of old age. Related forms. The Rhizopoda or pseudopodial Protozoa. *Arcella. Difflugia.* The "sun-animalcule." The Foraminifera. The Radiolaria.. 158

CHAPTER XIII.

UNICELLULAR ANIMALS (Continued).

B. INFUSORIA.

General account. Habitat. The "slipper-animalcule." The "bell-animalcule." *Paramœcium.* Its form, structure, and habits. Cytoplasm; trichocysts; vacuoles; nuclei; mouth; œsophagus; anal spot. The encysted state. Reproduction by agamogenesis; by conjugation; amphimixis. *Vorticella.* Its form, structure, etc. Its reproduction by fission, endogenous division, and conjugation. Microgamete and macrogamete. Related forms. *Euglena; Zoothamnion; Carchesium; Epistylis;* etc. Physiology of the Infusoria. Herbivorous, carnivorous, and omnivorous infusoria. Analogy with higher forms. The problem of chlorophyll in animals. Symbiosis. Vegetating animals. The claim of unicellular animals to be regarded as unicellular "organisms"; organs in the cell; etc.. 168

CHAPTER XIV.

UNICELLULAR PLANTS.

A. PROTOCOCCUS.

General account. Habitat. Morphology. Structure. Motile and non-motile states. Reproduction by fission. Cell-aggregates. Physiology. Income and outgo. The making of starch from inorganic matters. The fundamental physiological properties of protoplasm as displayed by plants Comparison of *Protococcus* with *Amœba*, and chlorophyll-bearing plants in general with animals in general. Other unicellular chlorophyll-bearing plants; diatoms; desmids; *Chroöcoccus; Glæocapsa;* etc.. 178

CHAPTER XV.

UNICELLULAR PLANTS (Continued).

B. YEAST.

General account. Wild yeast and domesticated yeast. Microscopical examination of a yeast-cake. Morphology of the yeast cell. Cytoplasm and nucleus. Reproduction by budding and by spores. Physiology. Yeast and the environment. Dried yeast. Income. Metabolism. Outgo. The minimal nutrients of yeast compared with those of *Protococcus* and *Amœba*. Why yeast is regarded as a plant. Top yeast Bottom yeast. Wild yeasts. Red yeast. Fermentation and ferments. Unicellular plants not necessarily at the bottom of the scale of life; etc 184

CHAPTER XVI.

UNICELLULAR PLANTS (Continued).

C. BACTERIA.

The smallest, most numerous, and most ubiquitous of known living things. Their abundance in earth, air, milk, water, etc. Comparison of their work in soils with that of earthworms. Parasitic and saprophytic bacteria. Their botanical position. Sanitary and economic importance. Morphology Structure. Cytoplasm and nucleus. Cilia. Their size. Swarming and the resting stages. Reproduction, Endospores. Arthrospores. Physiology. Income. Metabolism. Outgo. Ferments. Fermentation. Putrefaction. Disease. One species capable of living upon inorganic matter. Related forms Why bacteria are regarded as plants. The relations of bacteria to temperature, moisture, poisons, etc. Sterilization, Pasteurizing, disinfection, filtration, etc..................................... 192

CHAPTER XVII

A HAY INFUSION.

General account. Results of microscopical examination. Turbidity. Odor. Color. Constituents. The scene of important physical, chemical, and biological phenomena. Previous history of the hay and the water. Effect of bringing them together. Causes of turbidity, color, odor, etc. Aerobic and anaerobic bacteria thrive. Infusoria multiply and devour them. Carnivorous infusoria attack the herbivorous. The struggle for existence. Hay a green plant and the source of food. Quiet finally supervenes. How nutritive equilibrium may be preserved or disturbed. The hay-infusion an epitome of the living world.. 201

APPENDIX.

SUGGESTIONS FOR LABORATORY STUDIES AND DEMONSTRATIONS.

Books for the laboratory. Time required for General Biology 205
Special suggestions for laboratory work, etc., upon the subjects treated in the several chapters as outlined above, viz.:

- Chapter I. Introductory 205
- II. Structures of Living Organisms..................... 206
- III. Protoplasm and the Cell 207
- IV.-VIII. The Earthworm.............................. 210
- IX.-XI. The Fern...................................... 213
- XII. Amoeba.. 216
- XIII. Infusoria. 217
- XIV. Protococcus...................................... 220
- XV. Yeast... 221
- XVI. Bacteria... 222
- XVII. A Hay Infusion.................................. 223

INSTRUMENTS AND UTENSILS 220
REAGENTS AND TECHNICAL METHODS 221

INDEX... 227

GENERAL BIOLOGY.

CHAPTER I.

INTRODUCTORY.

We know from common experience that all material things are either dead or alive, or, more accurately, that all matter is either lifeless or living; and so far as we know, life exists only as a manifestation of living matter. Living matter and lifeless matter are everywhere totally distinct, though often closely associated. The most careful studies have on the whole rendered the distinction more clear and striking, and have demonstrated that living matter never arises spontaneously from lifeless matter, but only through the immediate influence of living matter already existing. And so, whatever may have been the case at an earlier period of the earth's history, we are justified in regarding the present line between living and lifeless as one of the most clearly defined and important of natural boundaries.

The Contrast between Living Matter and Lifeless Matter is made the ground for a division of the natural sciences into two great groups, viz.: the **Biological Sciences** and the **Physical Sciences**, dealing respectively with living matter and lifeless matter. The biological sciences (p. 7) are known collectively as **Biology** (βίος, *life*; λόγος, *a discourse*), which is therefore often defined as the science of life, or of living things, or of living matter. But living matter, so far as we know, is only ordinary matter which has entered into a peculiar state or condition.

And hence biology is more precisely defined as *the science which treats of matter in the living state.*

The Relationship between Living and Lifeless Matter. Although living matter and lifeless matter present this remarkable contrast to one another, they are most intimately related, as a moment's reflection will show. The living substance of the human body, or of any animal or plant, is only the transformed lifeless matter of the food which has been taken into the body and has there assumed, for a time, the living state. Lifeless matter in the shape of food is continually streaming into all living things on the one hand and passing out again as waste on the other. In its journey through the organism some of this matter enters into the living state and lingers for a time as part of the body substance. But sooner or later it dies, and is then for the most part cast out of the body (though a part may be retained within it, either as an accumulation of waste material, or to serve some useful purpose). Matter may thus pass from the lifeless into the living state and back again to the lifeless, over and over in never-ending cycles. A living plant or animal is like a fountain or a flame into which, and out of which, matter is constantly streaming, while the fountain or the flame maintains its characteristic form and individuality. It is "nothing but the constant form of a similar turmoil of material molecules, which are constantly flowing into the organism on the one side and streaming out on the other. . . . It is a sort of focus to which certain material particles converge, in which they move for a time, and from which they are afterward expelled in new combinations. The parallel between a whirlpool in a stream and a living being, which has often been drawn, is as just as it is striking. The whirlpool is permanent, but the particles of water which constitute it are incessantly changing. Those which enter it on the one side are whirled around and temporarily constitute a part of its individuality; and as they leave it on the other side, their places are made good by newcomers." (Huxley.)

How then is living matter different from lifeless matter? The question cannot be fully answered by chemical analysis, for the reason that this process necessarily kills living matter, and the results therefore teach us little of the chemical conditions existing in the matter when alive. Analyses, nevertheless, bring to light several highly important facts. It is likely that living matter is a tolerably definite compound of a number of the chemical elements, and it is probably too low an estimate to say that at least six elements must unite in order that life may exist. Moreover, only a very few out of all the elements are able, under any circumstances, to form this living partnership.

The most significant fact, however, is that there is no loss of weight when living matter is killed. The total weight of the lifeless products is exactly equal to the weight of the living substance analyzed, and if anything has escaped at death it is imponderable, and, having no weight, is not material. It follows that living matter contains no material substance peculiar to itself, and that every element found in living matter may be found also, under other circumstances, in lifeless matter.

Considerations like these lead us to recognize a fundamental fact, namely, that the terms living and lifeless designate two different STATES OR CONDITIONS of matter. We do not know, at present, what causes this difference of condition. But so far as the evidence shows, the living state is never assumed except under the influence of antecedent living matter, which, so to speak, infects lifeless matter and in some way causes it to assume the living state.

Distinctive Properties of Living Matter. Those properties of living matter which, taken together, distinguish it absolutely from every form of lifeless matter, are:

1. Its chemical composition.
2. Its power of waste and repair, and of growth.
3. Its power of reproduction.

Living matter invariably contains substances known as **proteids**, which are believed to constitute its essential material basis (see p. 33). Proteids are complex compounds of Carbon, Oxygen, Hydrogen, Nitrogen, Sulphur, and (in some cases at any rate) Phosphorus.

It has been frequently pointed out that each of these six elements is remarkable in some way: oxygen, for its vigorous combining powers; nitrogen, for its chemical inertia; hydrogen, for its great molecular mobility; carbon, sulphur, and phosphorus, for their allotropic properties, etc. All of these peculiarities may be shown to be of significance when considered as attributes of living matter. (See Herbert Spencer, *Principles of Biology*, vol. i.)

It is not, however, the mere presence of proteids which is characteristic of living matter. White-of-egg (albumen) contains an abundance of a typical proteid and yet is absolutely lifeless. Living matter does not simply contain proteids, but has the *power to manufacture them* out of other substances; and this is a property of living matter exclusively.

The waste and repair of living matter are equally characteristic. The living substance continually wastes away by a kind of internal combustion, but continually repairs the waste. Moreover, the growth of living things is of a characteristic kind, differing absolutely from the so-called growth of lifeless things. Crystals and other lifeless bodies grow, if at all, by *accretion*, or the addition of new particles to the outside. Living matter grows from within by *intussusception*, or the taking-in of new particles, and fitting them into the interstices between those already present, throughout the whole mass. And, lastly, living matter not only thus repairs its own waste, but also gives rise by reproduction to new masses of living matter which, becoming detached from the parent mass, enter forthwith upon an independent existence.

We may perceive how extraordinary these properties are by supposing a locomotive engine to possess like powers: to carry on a process of self-repair in order to compensate for wear; to grow and increase in size, detaching from itself at intervals pieces of brass or iron endowed with the power of growing up step by step into other locomotives capable of running themselves, and of reproducing new locomotives in their turn. Precisely these things are done by every living thing, and nothing like them takes place in the lifeless world.

Huxley has given the best statement extant of the distinctive properties of living matter, as follows:

"1. Its *chemical composition*—containing, as it invariably does, one or more forms of a complex compound of carbon, hydrogen, oxygen, and nitrogen, the so-called protein (which has never yet been obtained except as a product of living bodies), united with a large proportion of water, and forming the chief constituent of a substance which, in its primary unmodified state, is known as protoplasm.

"2. Its *universal disintegration and waste by oxidation, and its concomitant reintegration by the intussusception of new matter*. A process of waste resulting from the decomposition of the molecules of the protoplasm in virtue of which they break up into more highly oxidated products, which cease to form any part of the living body, is a constant concomitant of life. There is reason to believe that carbonic acid is always one of these waste products, while the others contain the remainder of the carbon, the nitrogen, the hydrogen, and the other elements which may enter into the composition of the protoplasm.

"The new matter taken in to make good this constant loss is either a ready-formed protoplasmic material, supplied by some other living being, or it consists of the elements of protoplasm, united together in simpler combinations, which constantly have to be built up into protoplasm by the agency of the living matter itself. In either case, the addition of molecules to those which already existed takes place, not at the surface of the living mass, but by interposition between the existing molecules of the latter. If the processes of disintegration and of reconstruction which characterize life balance one another, the size of the mass of living matter remains stationary, while if the reconstructive process is the more rapid, the living body *grows*. But the increase of size which constitutes growth is the result of a process of molecular intussusception, and therefore differs altogether from the process of growth by accretion, which may be observed in crystals, and is effected purely by the external addition of new matter; so that, in the well-known aphorism of Linnæus, the word 'grow' as applied to stones signifies a totally different process from what is called 'growth' in plants and animals.

"3. Its *tendency to undergo cyclical changes*. In the ordinary course of nature, all living matter proceeds from pre-existing living matter, a portion of the latter being detached and acquiring an independent existence. The new form takes on the characters of that from which it arose; exhibits the same power of propagating itself by means of an offshoot; and, sooner or later, like its predecessor, ceases to live, and is resolved into more highly oxidated compounds of its elements.

"Thus an individual living body is not only constantly changing its substance, but its size and form are undergoing continual modifications, the end of which is the death and decay of that individual; the continuation of the kind being secured by the detachment of portions which tend to run through the same cycle of forms as the parent. No forms of matter which are either not living or have not been derived from living matter exhibit these three properties, nor any approach to the remarkable phenomena defined under the second and third heads." (*Encyclopædia Britannica*, 9th ed., art. "Biology," vol. iii. p. 679.)

For the purposes of biological study life must be regarded as a property of a certain kind of compounded matter. But we are forced to regard the properties of compounds as the resultants of the properties of their constituent elements, even though we cannot well imagine how such a relation exists; and so in the long-run we have to fall back upon the properties of carbon, hydrogen, nitrogen, oxygen, etc., for the properties of living matter.

Scope of Biology. The Biological Sciences. It follows from the broad definition given to Biology that this science includes the study of whatever pertains to living matter or to living things. It considers the forms, structures, and functions of living things in health and in disease; their habits, actions, modes of nutrition; their surroundings and distribution in space and time, their relations to the lifeless world and to one another, their sensations, mental processes, and social relations, their origin and their fate, and many other topics. It includes both zoölogy and botany, and deals with the phenomena of animal and vegetal life not only separately, but in their relations to one another. It includes the medical sciences and vegetal pathology.

The field covered by biology as thus understood is so wide as to necessitate a subdivision of the subject into a number of principal branches which are usually assigned the rank of distinct sciences. These are arranged in a tabular view on p. 7. The table shows two different ways of regarding the main subject, according as the table is read from left to right or vice versa. Under the more usual arrangement biology is primarily divided into zoölogy and botany, according as animals or plants, respectively, form the subject of study. Such a division has the great advantage of practical convenience since, as a matter of fact, most biologists devote their attention mainly either to plants alone or to animals alone. From a scientific point of view, however, a better subdivision is into *Morphology* (μορφή, *form;* λόγος, *a discourse*) and *Physiology* (φύσις, *nature;* λόγος, *a discourse*). The former is based upon the facts of form, structure, and arrangement, and is essentially statical: the latter upon those of action or function, and is essentially dynamical. But morphology and physiology are so intimately related that it is impossible to separate either subject absolutely from the other.

Besides the sub-sciences given in the table a distinct branch called *Ætiology* is often recognized, having for its object the investigation of the causes of biological phenomena. But the scientific study of every phenomenon has for its ultimate object the discovery of its cause. Ætiology is therefore inseparable from

Biology.
The science of all living things; i.e., of matter in the living state.

Morphology.
The science of form, structure, etc. Essentially statical.

Anatomy.
The science of structure; the term being usually applied to the coarser and more obvious composition of plants or animals.

Histology.
Microscopic anatomy. The ultimate optical analysis of structure by the aid of the microscope; separated from anatomy only as a matter of convenience.

Taxonomy or Classification.
The classification of living things. Based chiefly on phenomena of structure.

Distribution.
Considers the position of living things in space and time, their distribution over the present face of the earth and their distribution and succession at former periods, as displayed in fossil remains.

Physiology.
The science of action or function. Essentially dynamical.

Embryology.
The science of development from the germ. Includes many mixed problems pertaining both to morphology and physiology. At present largely morphological.

Physiology.
The special science of the functions of the individual in health and in disease; hence including *Pathology*.

Psychology.
The science of mental phenomena.

Sociology.
The science of social life, i.e., the life of communities, whether of men or of lower animals.

Biology.
The science of all living things; i.e., of matter in the living state.

Botany.
The science of vegetal living matter or plants.

Zoölogy.
The science of animal living matter or animals.

any of the several branches of biology and need not be assigned an independent place.

Psychology and *Sociology* are not yet generally admitted to constitute branches of biology, and it is customary and convenient to set them apart from it. The establishment of the theory of evolution has clearly shown, however, that the study of these sciences is inseparable from that of biology in the ordinary sense. The instincts and other mental actions of the lower animals are as truly subjects of psychological as of physiological inquiry; the complex social life of such animal communities as we find, for instance, among the bees and ants are no less truly problems of Sociology.

It will be observed that in the scheme morphology and physiology overlap; that is, there are certain biological sciences in which the study of structure and of action cannot be separated. This is especially true of embryology, which considers the successive stages of embryonic structure and also the modes of action by which they are produced. And finally it must not be forgotten that any particular arrangement of the biological sciences must be in the main a matter of convenience only; for it is impossible to study any one order of phenomena in complete isolation from all others.

The term **General Biology** does not designate a particular member of the group of biological sciences, but is only a convenient phrase, which has come into use for the general introductory study of biology. It bears precisely the same relation to biology that general chemistry bears to chemistry or general physics bears to physics. It includes an examination of the general properties of living matter as revealed in the structures and actions of particular living things, and may serve as a basis for subsequent study of more special branches of the science. It deals with the broad characteristic phenomena and laws of life as illustrated by the thorough comparative study of a series of plants and animals taken as representative types; but in this study the student should never lose sight of the fact that all the varied phenomena which may come under his observation are in the last analysis due to *the properties of matter in the living state*, and that this matter and these properties are the real goal of the study.

CHAPTER II.

THE STRUCTURE OF LIVING THINGS. ORGANISMS.

Lifeless things occur in masses of the most various sizes and forms, and may differ widely in structure and chemical composition. Living things, on the other hand, occur only in relatively small masses, of which perhaps the largest are, among plants, the great trees of California and, among animals, the whales; while the smallest are the micro-organisms or bacteria. Moreover, the individual masses in which living things occur possess a peculiar and characteristic structure and chemical composition which have caused them to be known as *organisms*, and their substance as *organic*. All organisms are built up to a remarkable extent in the same way and of the same materials,

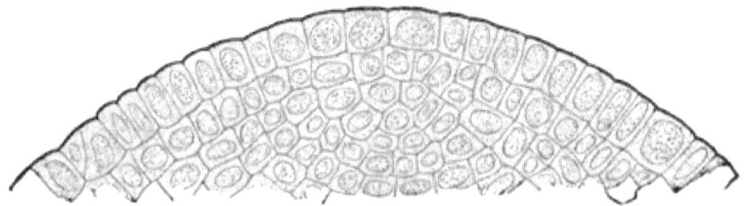

Fig. 1. (After Sachs.) Longitudinal section through the growing apex of a young pine-shoot. The dotted portion represents the protoplasm, the narrow lines being the partition-walls composed of cellulose ($C_6H_{10}O_5$). (Highly magnified.)

and we may conveniently begin a study of living things with the larger and more complex forms, which exhibit most clearly those structural peculiarities to which we have referred.

Organisms composed of Organs. Functions. It is characteristic of any living body—for example, a rabbit or a geranium—that it is composed of unlike parts, having a structure which enables them to perform various operations essential or accessory to the life of the whole. The plant has stem, roots, branches, leaves, stamens, pistil, seeds, etc.; the animal has externally head, trunk, limbs, eyes, ears, etc., and internally stomach, intestines, liver, lungs, heart, brain, and many other parts of

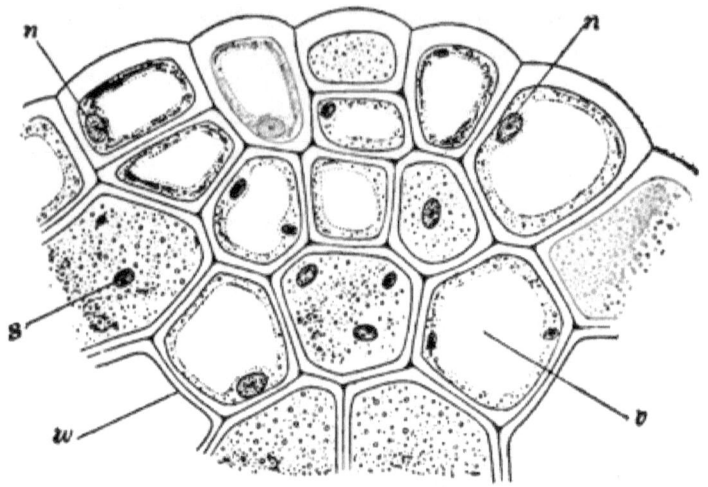

FIG. 2.—Cross-section through part of the young leaf of a fern (*Pteris aquilina*), showing thick-walled cells; most of the walls are double. The granular substance is protoplasm. Most of the cells contain a large central cavity (vacuole) filled with sap, the protoplasm having been reduced to a thin layer inside the partitions. Nuclei are shown in some of the cells, and lifeless grains of starch in others: n, nuclei; s, starch; v, vacuole; w, double partition-wall. (× 500.)

the most diverse structure. These parts are known as *organs*, and the living body, because it possesses them, is called an *organism*.

The word organism, as here used, applies best to the higher animals and plants. It will be seen in the sequel that there are forms of life so simple that organs as here defined can scarcely be distinguished. Such living things are nevertheless really organisms because they possess parts analogous in function to the well-defined organs of higher form. (See p. 157.)

Since organisms are composed of unlike parts, they are said to be heterogeneous in structure. They are also heterogeneous in action, the different organs performing different operations called *functions*. For instance, it is the function of the stomach to digest food, of the heart to pump the blood into the vessels, of the kidneys to excrete waste matters from the blood, and of the brain to direct the functions of other organs. A similar diversity of functions exists in plants. The roots hold the plant fast and absorb various substances from the soil; the stem supports the leaves and flowers and conducts the sap; the leaves absorb and elaborate portions of the food; and the reproductive organs of the flower serve to form and bring to maturity seeds destined to give rise to a new generation.

Heterogeneity of the kind just indicated, accompanied by *a division of labor* among the parts, is one of the most characteristic features of living things, and is not known in any mass of lifeless matter, however large and complex.

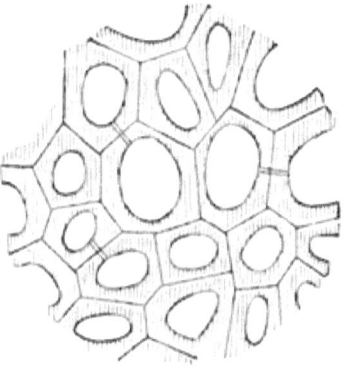

FIG. 3. (After Sachs.) Cross-section through a group of dead, thick-walled wood-cells from the stem of maize. The cells contain only air or water. (Highly magnified.)

Organs composed of Tissues. Differentiation. In the next place, it is to be observed that the organs also, when fully formed, are not homogeneous, but are in turn made up of different parts. The human hand is an organ which consists of many parts, differing widely in structure and function. On the outside are the skin, the hairs, the nails; inside are bones, muscles, tendons, ligaments, blood-vessels, and nerves. The leaf of a plant is an organ consisting of a woody framework (the "veins") which supports a green pulp, the whole being covered on the outside by a delicate transparent skin. In like manner every organ of the higher plants or animals may be resolved into different parts, and these are known as *tissues*. The tissues of fully formed organs are often very different from one another, as in the cases just mentioned; that is, they are well *differentiated;* but frequently in adult organs, and always in those which are sufficiently young, the tissues shade gradually into one another, so that no definite line can be drawn between them. In such cases they are said to be less differentiated. For example, in the full-grown leaf of a plant the woody framework, the green cells, and the skin exist as three plainly different tissues. But in younger leaves these same tissues are less different, and in very young leaves, still in the bud, there are no visible differences and the whole organ is very nearly homogeneous. In this case the tissues are *undifferentiated*, though potentially capable of differentiation. In the same way, the tissues of the embry-

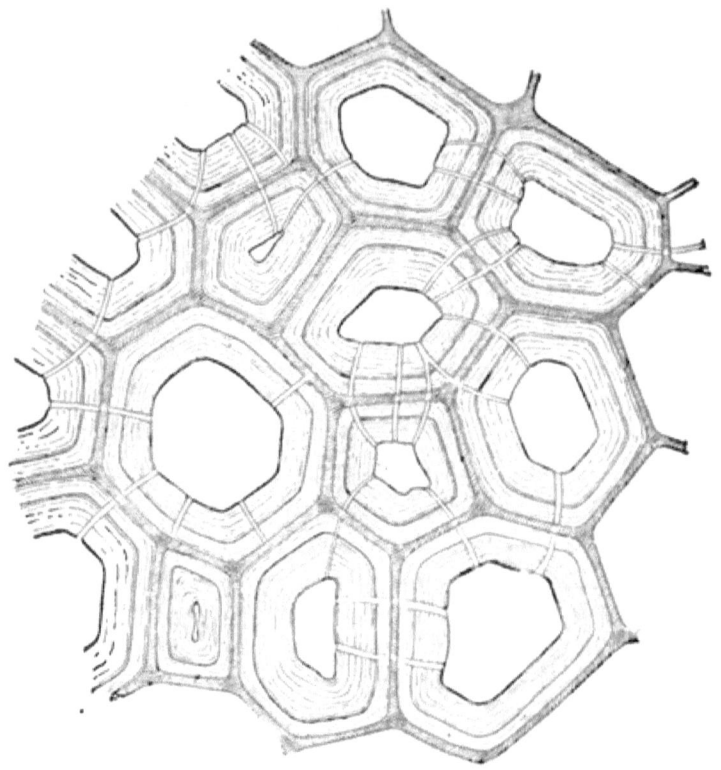

FIG. 4.—Cross-section through dead wood-like cells from the underground stem of a fern (*Pteris aquilina*). The walls are uncommonly thick and the protoplasm has disappeared. The channels shown served in life to keep the cells in vital connection. (× 50.)

onic human hand are imperfectly differentiated, and at a very early stage are undifferentiated.

Tissues composed of Cells. Finally, microscopical examination shows every tissue to be composed of minute parts known as *cells*, which are nearly or quite similar to one another throughout the whole tissue, and form the ultimate units into which the tissues and organs, and hence the whole organism, become more or less perfectly divided, somewhat as a nation is divided into states and these into counties and townships.

It will be shown beyond that these ultimate units or cells possess everywhere the same fundamental structure; but they differ immensely in form, size, and mode of action, not only in different animals and plants, but even in different parts of the same individual. As a rule, the cells of any given tissue are closely similar one to another and are devoted to the same function, but differ from those of other tissues in form, size, arrangement, and especially in function. Indeed, the differences between tissues are merely the outcome of the differences between the cells composing them. The skin of the hand differs in appearance and uses from the muscle which it covers, because skin-cells differ from muscle-cells in form, size, color, function, etc. Hence a tissue may be defined as *a group of similar cells having a similar function.** As a rule, each organ consists of several such groups of cells or tissues, but, as stated above, young organs are nearly or quite homogeneous; that is, all of the cells are nearly or quite alike. It is only when the organ grows older that the cells become different and arrange themselves in different groups,—a process known as the *differentiation of the tissues*. In the case of some organs—for instance the leaf of a moss—the cells remain permanently nearly alike, somewhat as in the embryonic condition, and the whole organ consists of a single tissue.

What has been said thus far applies only to higher plants and animals. But it is an interesting and suggestive fact that there are also innumerable isolated cells, both vegetal and animal, which are able to carry on an independent existence as one-celled plants or animals. Physiologically these must certainly be regarded as individuals; but it is no less certain that they are equivalent, morphologically, to the constituent cells of ordinary many-celled organisms. It will appear hereafter that the study of such unicellular organisms forms the logical groundwork of all biological science. (See p. 157.)

Since organisms may be resolved successively into organs, tissues and cells, it is evident that cells must contain living matter. And a cell may be defined as a *small mass of living matter either living apart or forming one of the ultimate units*

---

* Tissues frequently contain matters deposited between cells; but these have usually been directly derived from the cells, and vary as the cells vary.

of an organism. The cell is an "*organic individual of the first order.*" (Lang.)

Living and Lifeless Matter in the Living Organism. Since our own bodies and those of lower animals and of plants are composed of matter, it may be supposed, from what has been said in the last chapter, that they are composed of living matter. This, however, is true only in part. It is strictly true that every plant or animal contains living matter, but a little reflection will show that it contains lifeless matter also. In the human body lifeless matter is found in the hairs, the ends of the nails, and the outer layers of the skin,—structures which are not simply devoid of feeling, as every one knows them to be, but are really lifeless in every sense, although forming part of a living body. Nor is lifeless matter confined to the exterior of the body. The mineral matter of the bones is not alive; and this is true, though less obviously, of many other parts, such as the liquid basis or plasma of the blood, the fat (which is never wholly absent), and various other forms of matter occurring in many parts of the body.

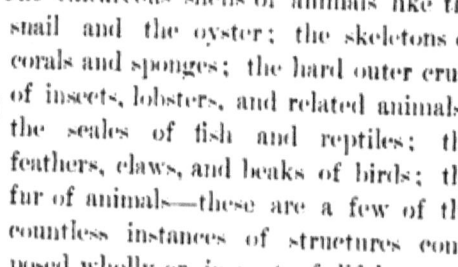

FIG. 5. (After Ranvier.) Muscle-cells. *A*, from the intestine of a dog, in cross-section; *B*, single isolated cell, from the intestine of a rabbit, viewed from the side. × 320.)

In lower animals examples of this truth occur on every hand. The calcareous shells of animals like the snail and the oyster; the skeletons of corals and sponges; the hard outer crust of insects, lobsters, and related animals; the scales of fish and reptiles; the feathers, claws, and beaks of birds; the fur of animals—these are a few of the countless instances of structures composed wholly or in part of lifeless matter, which nevertheless enter into the composition of living animals.

Among plants like facts are even more conspicuous. No one can doubt that the outer bark of an oak is devoid of life. The heart-wood of a tree is entirely dead, and even in the so-called live wood, through which the sap flows, not only is the solid part of the wood lifeless, but also the sap itself.

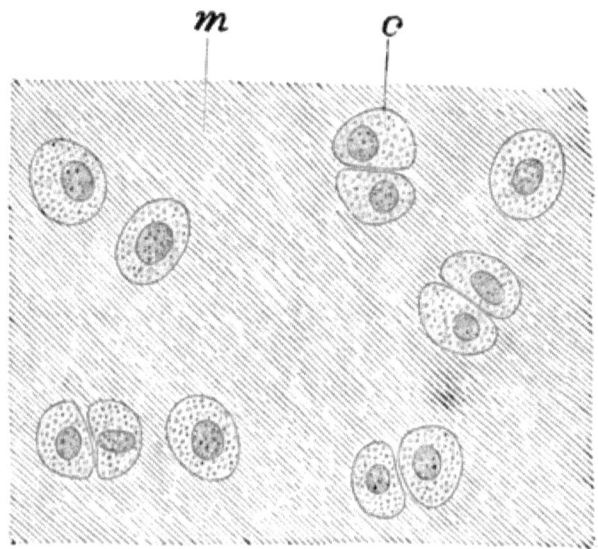

FIG. 6. (After Schäfer.)—Human cartilage (from head of metatarsal bone). *c*, cells; *m*, lifeless matrix. (× 600.)

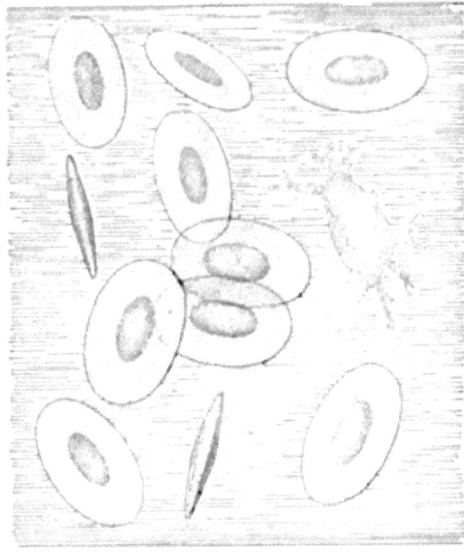

FIG. 7. (Modified from Ranvier.)—Blood of frog, showing two forms of cells (corpuscles), one flattened and oval, one branched, floating in the lifeless plasma. (× 650.)

Lifeless Matter in the Living Tissues. In the tissues the living cells are seldom in contact one with another, but are more or less completely separated by partitions of lifeless matter. This may be seen in a section through some rapidly growing organ like a young shoot (Fig. 1). The whole mass is formed of nearly similar, closely crowded units or cells separated by very narrow partitions. Each cell consists of a mass of granular, viscid, living substance known as *protoplasm*, and a more solid, rounded body, the *nucleus*.

In such a group of cells no tissues can be distinguished; or, rather, the whole mass consists of a single tissue (meristem), which is almost entirely composed of living matter (protoplasm). In older tissues the partitions often increase in thickness, as shown in Fig. 2. In every case *the partitions are composed of lifeless matter which has been manufactured and deposited by the living protoplasm constituting the bodies of the cells*. In still older parts of the plant certain of the lifeless walls may become extremely thick, the protoplasm entirely disappears, and the whole tissue (wood) consists of lifeless matter enclosing spaces filled with air or water (Figs. 3 and 4).

Among animals analogous cases are common. The muscles of the small intestine, for instance, (Fig. 5,) consist of bundles of elongated cells (*fibres*) each of which is composed of living matter surrounded by a very thin covering (*sheath*) of lifeless matter. In cartilage or gristle, which covers the ends of many bones (Fig. 6), the oval cells are very widely separated by the deposition between them of large quantities of solid lifeless matter forming what is known as the *matrix*. In blood (Fig. 7) the flattened or irregular cells (*corpuscles*) are separated by a lifeless fluid (*plasma*) in which they float. In bone (Fig. 8) the cells

FIG. 8. (Modified from Schenk.)—Section of bone from the human femur showing the living branching bone-cells lying in the bony lifeless matrix. Diagramatic.

have a branching, irregular form, and are separated by solid calcareous matter which is unmistakably lifeless. These examples show that the lifeless matters of the body often occur in the form of deposits between living cells by which they have been produced. In all such cases the embryonic tissue consists at first of living cells in direct contact, or separated by only a very small quantity of lifeless matter. In later stages the cells may manufacture additional lifeless substance which appears in the form of firm partition-walls between the cells, or as a matrix, solid or liquid, in which the cells lie. When solid walls are present they are often perforated by narrow channels through which the protoplasmic cell-bodies remain in connection. (See Figs. 4, 8, and 50.)

Lifeless Matter within Living Cells. Equally important with the deposit of lifeless matter *between* cells is the formation of lifeless matter *within* cells, either (*a*) by the deposition of various substances in the protoplasm, or (*b*) by the direct transformation of the whole mass of protoplasm. Examples of the first kind are

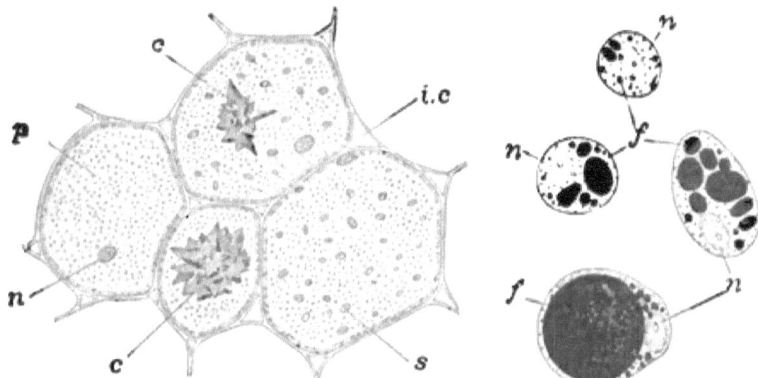

FIG. 9.—A group of cells from the stem of a geranium (*Pelargonium*), showing lifeless substances (starch and crystals) within the protoplasm. As in Fig. 2, each cell contains a large central vacuole, filled with sap; *c*, groups of crystals of calcium oxalate; *i.c.*, intercellular space; *n*, nucleus; *s*, granules of starch. (× 300.)

FIG. 10. (After Ranvier.)— Group of "adipose cells" from the tissue beneath the skin ("subcutaneous connective tissue") of an embryo calf, showing drops of fat in the protoplasm. *f*, fat-drops (black); *n*, nuclei (× 550.)

mineral crystals (Fig. 9), grains of starch (Fig. 9), drops of water, and many other substances found within the cells of plants. Among animals drops of fat (Fig. 10) and calcareous or siliceous deposits are similarly produced. Indeed, there is scarcely any limit to the number of lifeless substances which may thus appear within the cells both of plants and animals.

The second case is of less importance, though of common occurrence. A good example is found in the lining membrane of the œsophagus of the dog (Fig. 11), which like the human skin is almost entirely made up of closely crowded cells. Those

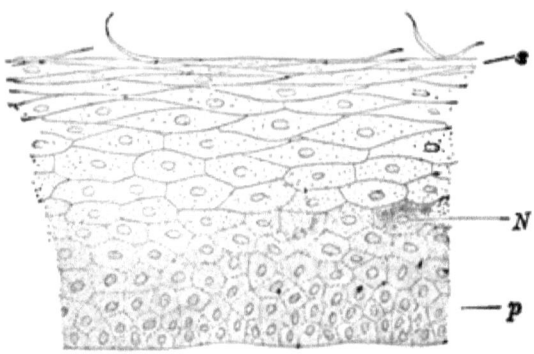

FIG. 11.—Section through the inner coat of the gullet of a dog, showing: p, living cells of the deeper layers; s, lifeless cells of the superficial layers; n, nucleus.

in the deepest part consist chiefly of living protoplasm very similar to that of the young pine shoot (compare Fig. 1). Above them the cells gradually become flattened until at the surface they have the form of flat scales. As the cells become flattened their substance changes. The protoplasm diminishes in quantity and dies; so that near the surface the cells are wholly dead, and finally fall off. In a similar manner are formed the lifeless parts of nails, claws, beaks, feathers, and many related structures. A hair is composed of cells essentially like those of the skin. At the root of the hair they are alive, but as they are pushed outwards by continued growth at the root, they are transformed bodily into a dead, horny substance forming the free portion of the hair. Feathers are only a complicated kind of hair and are formed in the same way.

It is a significant fact that the quantity of lifeless matter in the organism tends to increase with age. The very young plant or animal probably possesses a maximum proportion of protoplasm, and as life progresses lifeless matter gradually accumulates within or about it,—sometimes for support, as in tree-trunks and bony skeletons; sometimes for protection as in oyster- and snail-shells; sometimes apparently from sheer inability on the part of the protoplasm to get rid of it. Thus we see that youth is literally the period of life and vigor, and age the period of comparative lifelessness.

Summary. The bodies of higher animals and plants are subdivided into various parts (*organs*) having different structure and functions. These may be resolved into one or more *tissues*, each of which consists of a mass of similar *cells* (or their derivatives) having a similar function. The cells are small masses of living matter, or protoplasm, which deposit more or less lifeless matter either around (outside) them or within their substance. In the former case the protoplasm may continue to live, or it may die and be absorbed. In the latter case it may likewise live on for a time, or may die, either disappearing altogether or leaving behind a residue of lifeless matter.

The Organism as a Whole. Up to this point we have considered living organisms from an anatomical and analytical standpoint, and have observed their natural subdivisions into organs, tissues, and cells. We have now only to remark that these parts are mutually interdependent, and that the organism as a whole is greater than any of its parts. Precisely as a chronometer is superior to an aggregate of wheels and springs, so a living organism is superior in the solidarity of its parts to a mere aggregate of organs, tissues, and cells. We shall soon see that in the living body these have had a common ancestry and still stand in the closest relationship both in respect to structural continuity and community of interest.

CHAPTER III.

PROTOPLASM AND THE CELL.

It has been shown in the last chapter that life is inherent in a peculiar substance, *protoplasm*, occurring in definite masses or *cells*. In other words, protoplasm is the physical basis of life, and the cell is the ultimate visible structural unit. Protoplasm and the cell deserve therefore the most careful consideration; but because of the technical difficulties involved in their study only such characteristics as are either obvious or indispensable to the beginner will here be dwelt upon.

Historical Sketch. Organs and tissues are readily visible, but in order to resolve tissues into cells something more than the naked eye was necessary. The compound microscope came into use about 1650, and in 1665 the English botanist Robert Hooke announced that a familiar vegetal tissue, cork, is made up of "*little boxes or cells distinct from one another.*" Many other observers described similar cells in sections of wood and other vegetal tissues, and the word soon came into general use. It was not until 1838, however, and as a consequence of a most important improvement in the compound microscope, viz., the invention of the achromatic objective, that cellular structure came to be recognized as an invariable and fundamental characteristic of living bodies. At this time the botanist Schleiden brought forward proof that the higher plants do not simply contain cells but are wholly made up of them or their products; and about a year later the zoölogist Schwann demonstrated that the same is true of animals. This great generalization, known as the "*cell-theory*" *of Schleiden and Schwann*, laid the basis for all subsequent biological study. The cell-theory was at first developed upon a purely morphological basis. Its application to the phenomena of physiological action was for a time retarded by the misleading character of the term "cell." The word itself shows that cells were at first regarded as cavities (like the cells of a honeycomb or of a prison) surrounded by solid walls; and even Schleiden and Schwann had no accurate conception of their true nature. Soon after the promulgation of the cell-theory, however, it was shown that both the walls and the cavity might be wanting, and that therefore the remaining portion, namely, the protoplasm with its nucleus, must be the active and essential part. The cell was accordingly defined by Virchow and Max Schultze as "a mass of protoplasm surrounding a nucleus," and in this sense the word is used to-day.* The word cell became thereafter as inappropriate as it would be if applied to the honey within the honeycomb or to the living prisoner in a prison-cell. Nevertheless, by a curious conservatism, the term was and is retained to designate these structures whether occurring in masses, as segments of the plant or animal body, or leading independent lives as unicellular organisms.

Protoplasm was observed long before its significance was understood. The discovery of its essential identity in plants and animals and, ultimately, the general recognition of the extreme importance of the *rôle* which it everywhere plays, must be reckoned as one of the greatest scientific achievements of this century. It was Dujardin who in 1835 first distinctly called attention to the importance of the "primary animal substance" or "sarcode" which forms the bodies of the simplest animals. Without clearly recognizing this substance as the seat of life, or using the word protoplasm, he nevertheless described it as endowed with the powers of spontaneous movement and contractility. The word protoplasm ($\pi\rho\hat{\omega}\tau o \varsigma$, first; $\pi\lambda\acute{a}\sigma\mu a$, form) was apparently first used for animal substance by Purkinje in 1839–40, and next by H. von Mohl, in 1846, to designate the granular viscid substance occurring in plant-cells, although both workers were ignorant of its full significance. In 1850 Cohn definitely maintained not only that animal sarcode and vegetal protoplasm were essentially of the same nature, but also that this substance is the real seat of vitality and hence to be regarded as the physical basis of life. To Max Schultze

---

* It is possible that in some of the lowest and simplest organisms even the nucleus may be wanting as a distinctly differentiated body. See p. 193.

(1860) is generally assigned the credit of having finally placed this conclusion upon a secure basis; and by him the meaning of the word **Protoplasm** was so extended as to include all **living matter**, whether animal or vegetal. In this sense the word is now universally employed.

Appearance and Structure. Protoplasm and cells differ greatly in appearance in different plants and animals, as well as in different parts and different stages of development of the same individual. The appearance of protoplasm and the constitution of the cell are as a rule most easily made out in very young structures, such as the eggs of some animals or in the cells of young vegetal shoots. The egg of the starfish, for example, (Fig. 12), is a single isolated cell of nearly typical form and structure. It is a minute, nearly spherical body ($\frac{1}{50}$ inch diameter) in which three parts may be distinguished, viz.: (1) the *cell-body*, which forms the bulk of the cell; (2) the *nucleus*, a rounded vesicular body suspended in the cell-body; (3) the *membrane* or *cell-wall*, which immediately surrounds the cell-body. Of these three, the nucleus and cell-body are mainly composed of protoplasm, while the membrane is a lifeless deposit upon the exterior. The protoplasm of the cell-body is generally called cell-plasm, or *cytoplasm*, that of the nucleus *nucleoplasm;* that is, the living matter of the cell is differentiated into two different but closely related forms of protoplasm, cytoplasm and nucleoplasm.

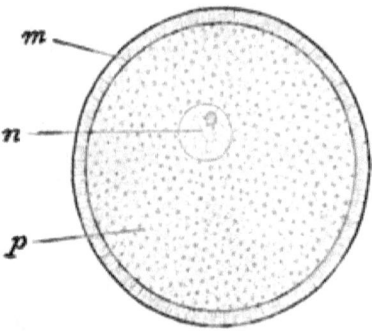

FIG. 12.—Slightly diagrammatic figure of the egg or ovum of a star-fish, showing the structure of a typical cell. *m*, membrane; *n*, nucleus; *p*, protoplasm (cytoplasm).

The Cytoplasm appears as a clear semifluid or viscid substance, containing numerous minute granules and of a watery appearance, though it shows no tendency to mix with water. Under very high powers of the microscope, especially after treatment with suitable reagents, the clear substance is found to have a definite structure, the precise nature of which is in dispute. By some observers it is described as a fibrous meshwork or reticulum, like a sponge; by others as more nearly like an emulsion or foam, consisting of a more solid framework enclosing innumerable minute separate spherical cavities filled with liquid; by others still as composed of unbranched threads running in all directions through a more liquid basis; but its real nature is still unknown.

It is evident that the visible structure of protoplasm gives no hint of its marvellous powers as the seat of vital action, and we are therefore compelled to infer that it is endowed with a chemical and molecular constitution extremely complex, and probably far exceeding in complexity that of any lifeless substance.

The Nucleus is a rounded body suspended in the cell-substance; it is distinguishable from the latter by its higher refractive power, and by the intense color it assumes when treated with staining fluids. It is surrounded by a very thin membrane, and consists internally of a clear substance (*achromatin*), through which extends an irregular network of fibres (*chromatin*). It is especially these fibres which are stained by dyes. In the

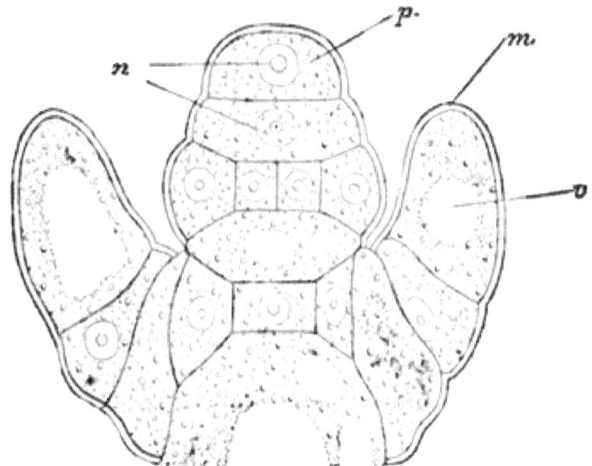

FIG. 13. (After Sachs.)—Young growing cells from the extreme tip of a stonewort (*Chara*). m, membrane; n, nuclei; p, protoplasm; v, vacuole filled with sap. (× 550.)

meshes of the network is suspended in many cases a second rounded body known as the *nucleolus*, which stains even more deeply than the network itself.

The Membrane or Wall of the cell forms a rather thick sac, composed of a soft, lifeless material closely surrounding the cell substance.*

As a second example we choose the growing point of a common water-plant (*Chara*), Fig. 13. This structure is composed of cells which are more or less angular in outline as a result of mutual pressure, but show otherwise an unmistakable similarity to the egg-cell just described. They differ mainly in the fact that the protoplasm of the larger cells contains rounded cavities, known as *vacuoles*, filled with sap (*v*); also in the chemical composition of the cell-walls (here consisting of "cellulose," a substance of rare occurrence among animals).

Origin of Cells and Genesis of the Body. The body of every higher plant or animal arises from a single germ-cell ("egg," "spore," etc.) more or less nearly similar to that of the starfish, described above, and originally forming a part of the parent body. The germ-cell, therefore, in spite of endless variations in detail, shows us the model after which all others are built; for it gives rise to all the cells of the body by a continued process of segmentation as follows:

The first step (Fig. 14) consists in the division of the egg into two similar halves, which differ from the original cell only in lacking membranes, both being surrounded by the membrane of the original cell. Each of the halves divides into two, making four in all; these again into two, making eight, and so on throughout the earlier part of the development. By this process (known as the cleavage or *segmentation* of the egg) the germ-cell gives rise successively to 2, 4, 8, 16, 32, 64, etc., descendants, forming a primitive body composed of a mass of nearly similar cells, out of which, by still further division and growth, the fully-formed body of the future animal is to be built up. These cells are only slightly modified, but differ in most animals from the typical germ-cell in having at first no surrounding membranes. The membrane of the original germ-cell meanwhile disappears.

---

* The word cell has been used in Chap. I and elsewhere to denote the living matter within the membrane, the latter being considered a product of the cell rather than an integral part of it. It is more usual to include the membrane in a definition of the cell, and as a matter of convenience it is so included here.

The embryonic body or *embryo* of every higher plant and animal is derived from the germ-cell by a process essentially like that just described, though both the form of the cells and the order of division are usually more or less irregular. In animals the cells

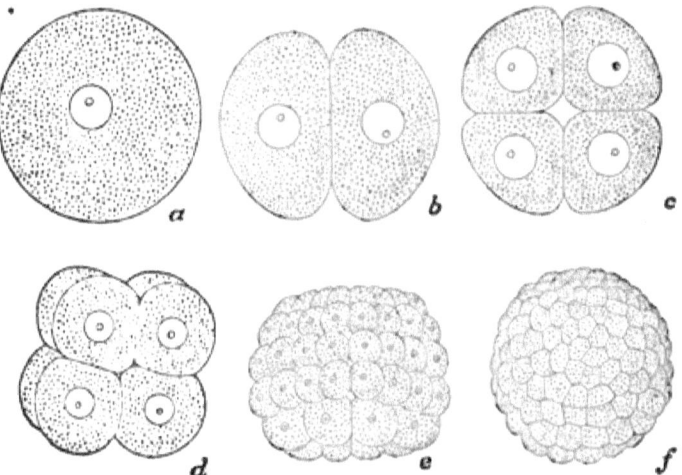

FIG. 14.—Cleavage or segmentation of an ovum, showing successive division of the germ-cell (*a*) into two (*b*), four (*c*), and eight (*d*). Later stages are shown at *e* and *f*. The first four figures are diagrammatic; *e* and *f* are after Hatschek's figures of the development of a very simple vertebrate (*Amphioxus*).

thus formed are usually naked at first, though they often acquire a membrane in later stages. Among plants, on the contrary, the cells usually possess membranes from the first, probably because their need for a firm outer support is greater than the need for free movement demanded by animals.*

Modification of the Embryonic Cells. Differentiation. The close similarity of the embryonic cells does not long persist. As development proceeds, the cells continually increasing in number by division become modified in different ways, or *differentiated*, to fit them for the many different kinds of work which they have to do. Those which are to become muscle-cells gradually assume an entirely different form and structure from those which are to become skin-cells; and the future nerve- or gland-cells take on still other forms and structures. The embryonic cells are gradually converted into the elements of the different tissues—this process being the *differentiation of the tissues* which has

---

* For a more precise account of cell-division see p. 83.

already been mentioned on p. 11—and are in this way enabled to effect a *physiological division of labor*.

The variations in form and structure which thus appear are endlessly diversified. Cells may assume almost any conceivable form, and there are even cells (e.g., *Amœba*, or the colorless corpuscles of the blood) which continually change their form from moment to moment. The variations in structure may involve any or all of the three characteristic parts of the typical cell, being at the same time accompanied by variations of form. It is easy to understand, therefore, how cells may vary endlessly in appearance, while conforming more or less closely to the same general type.

Meanwhile the protoplasm itself undergoes extensive alteration. Even in young cells, or in the germ-cell itself, it may contain an admixture of other substances, and these may entirely change their character or (as is especially common in plant-cells) may become more abundant as the cell grows older, taking the shape of fluid, solid, or even gaseous deposits. Common examples of such deposits are drops of water, oil, and resin, granules of pigment, starch, and solid proteid matters, and crystals of mineral substances like calcium oxalate, phosphate and carbonate, and silica. Bubbles of gas sometimes appear in the protoplasm, but this is exceptional. The living substance itself often changes in appearance as the cells become differentiated. The protoplasm of voluntary muscles (Fig. 15) is firm, clear, non-granular, highly refractive, and arranged in alternating bands or stripes of darker and lighter substance. In some cases (e.g., the outer portions of the skin, or of a hair, as explained in Chap. II) the modifications of the cell-substance becomes so great that both its physical and chemical constitution are entirely altered, and it is no longer protoplasm, but some form of lifeless matter.

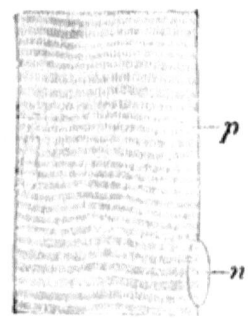

FIG. 15. (After Ranvier.)—Part of a single fibre of voluntary muscle from the leg of a rabbit. *p*, protoplasm; *n*, nucleus. (× 700.)

Protoplasm in Action. We may now briefly consider protoplasm from the dynamical or physiological point of view. We know that living things are the seat of active changes, which taken together constitute their life. In the last analysis these changes are undoubtedly chemical actions taking place in the protoplasm, which may or may not produce visible results. There is no doubt that extensive and probably very complex molecular actions go on in the protoplasm of young growing cells, though it may appear absolutely quiescent to the eye, even under a powerful microscope. In other cases, the chemical action produces perceptible changes in the protoplasm,—for instance, some form of motion,—just as the invisible chemical action in an electrical battery may be made to produce visible effects (light, locomotion, etc.) through the agency of an electrical machine.

A familiar instance of protoplasmic movement is the contraction of a muscle. This process is most likely a change of molecular arrangement, causing the muscle, while keeping its exact bulk, to change its form, the two ends being brought nearer together (Fig. 16). The visible change of form is here supposed to be due to an invisible change of molecular arrangement, and this in turn to be coincident with chemical action taking place in the living substance.

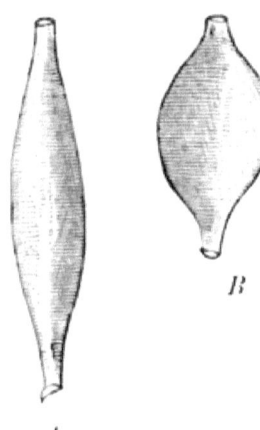

Fig. 16.—Change of form in a contracting muscle. *A*, muscle in the ordinary or extended state; *B*, the same muscle when contracted. (Diagram.)

A striking and beautiful example of movement in protoplasm occurs in the simple organism known as *Amœba* (Fig. 84, p. 159). The entire body of this animal consists of a mass of naked protoplasm enclosing a nucleus, or sometimes two; in other words, it is a single naked cell. The protoplasm of an active *Amœba* is in a state of ceaseless movement, contracting, expanding, flowing, and changing the form of the animal to such an extent that it is known as the "Proteus" animalcule. The whole movement is a kind of flux. A portion of the protoplasm flows out from the mass, making one or more prolongations (*pseudopods*) into which the remainder of the protoplasm finally passes, so that the whole body advances in the direction of the flow. If particles of food be met with, the protoplasm flows around them, and when they have been digested within the body, the protoplasm flows onward, leaving the refuse behind. Hour after hour and day after day this flowing may go on, and there is perhaps no more fascinating and suggestive spectacle known to the biologist. A similar change of form is exhibited by the colorless corpuscles of amphibian and other blood, in which it may be observed, though far less satisfactorily, if *Amœba* cannot be obtained. Among plants, protoplasmic movements of perhaps equal beauty may be observed. One of the simplest is known as the *rotation* of protoplasm, which may

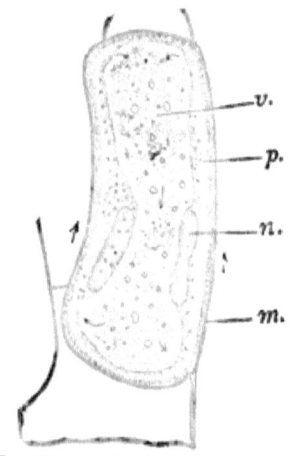
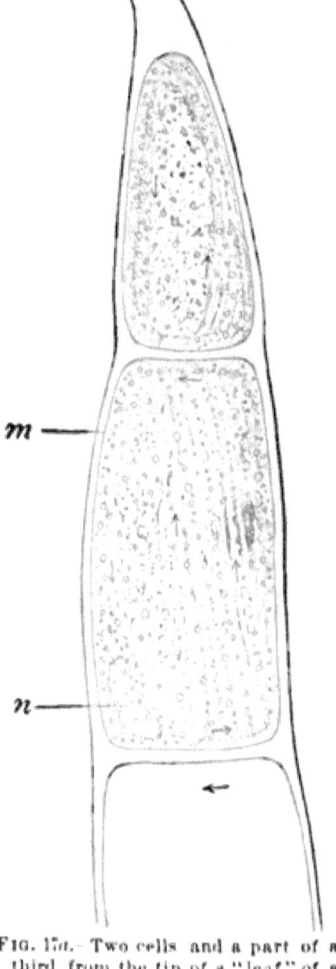

Fig. 17.—A cell of a stonewort (*Nitella*) showing the rotation of protoplasm; the arrows show the direction of the flow. *m*, membrane of the cell; *n*, nucleus, opposite to which is a second; *p*, protoplasm; *v*, large central vacuole filled with sap.

Fig. 17a.—Two cells and a part of a third from the tip of a "leaf" of a stonewort, showing rotation of the protoplasm in the direction of the arrows.

be studied to advantage in rather young cells of stoneworts (*Chara* or *Nitella*). These cells have the form of short or elongated cylinders which are often pointed at one end (Fig. 17). The protoplasm is surrounded by a delicate membrane which thus forms a sac enclosing the protoplasm. In very young cells the protoplasm entirely fills the sac; but as the cell grows older a drop of liquid appears near the centre of the mass and increases in size until the protoplasm is reduced to a thin layer (*primordial utricle*), lining the inner surface of the membrane (compare Fig. 2). In favorable cases the entire mass of protoplasm is seen to be flowing steadily around the inside of the sac, as indicated by the arrows in Fig. 17. It moves upwards on one side, downwards on the opposite side, and in opposite directions across the ends, forming an unbroken circuit. The flow is rendered more conspicuous by various granules and other lifeless masses floating in the protoplasm and by the large oval nucleus or nuclei, all of which are swept onward by the current in its ceaseless round. A similar rotation of protoplasm occurs in many other vegetal cells, one of the best examples being the leaf-cells of *Anacharis*.

A second and somewhat more intricate kind of movement in vegetal protoplasm is known as *circulation*. This differs from rotation chiefly in the fact that the protoplasm travels not only in a peripheral stream but also in strands which run across through the central space (vacuole) and thus form a loose network. Cir-

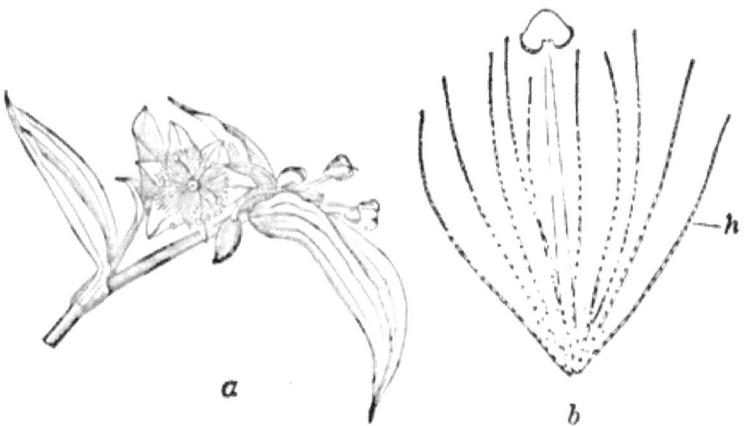

FIG. 18.—Flower-cluster (*a*) and single stamen (*b*) of a cultivated spiderwort (*Tradescantia*). *h*, hairs upon the stamen. *a*, slightly reduced; *b*, slightly enlarged

culation is well seen in cells composing the hairs of various plants, such as the common nettle (*Urtica*), the spiderwort (*Trades-*

cantia), the hollyhock (*Althæa*), and certain species of gourds (*Cucurbita*). It may be conveniently studied in the hairs upon the stamens of the cultivated spiderwort (*Tradescantia*). The flower of this plant is shown in Fig. 18, *a*, and one of the stamens with its hairs at *b*. Each hair consists of a single row

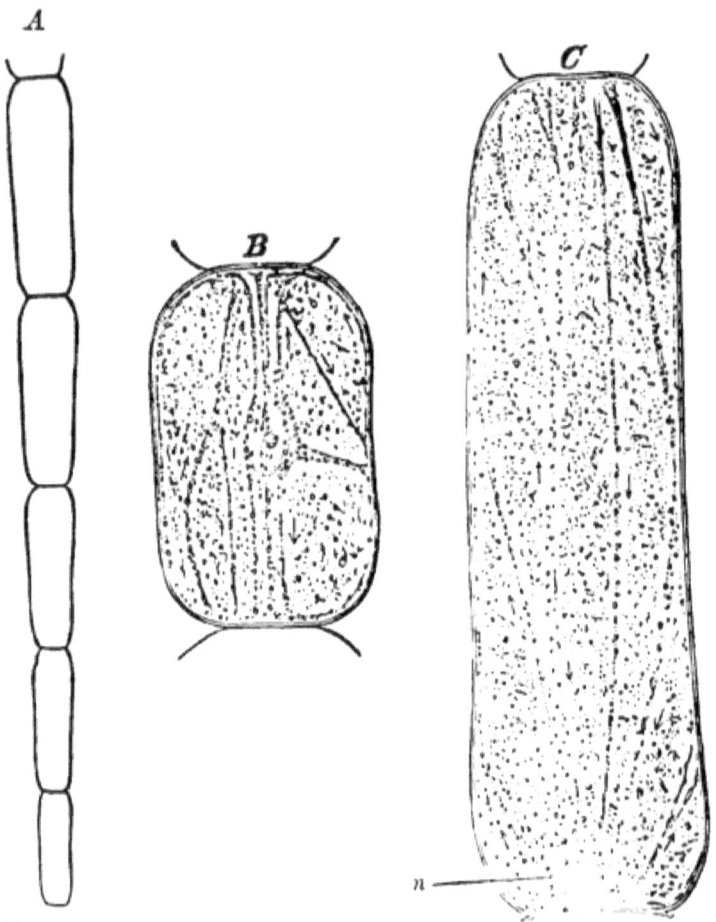

FIG. 19.—Enlarged cells of the hairs from the stamens of the spiderwort. *A*, five cells, somewhat enlarged, protoplasm not shown; *B* and *C*, cells much more enlarged, showing the circulation of protoplasm as indicated by the arrows; *n*, nucleus.

of elongated cells covered by delicate membranes and connected by their ends. As in *Nitella*, the protoplasm does not fill the cavity of the sac, but forms a thin lining (*primordial utricle*) on its inner face (Fig. 19). From this layer delicate threads of protoplasm reach into and pass through the central cavity, where they often branch and are connected together so as to form a very loose network. The nucleus (n) is embedded either in the peripheral layer or at some point in the network, and the threads of the latter always converge more or less regularly to it. In active cells currents continually flow to and fro throughout the whole mass of protoplasm. In the threads of the network granules are borne rapidly along, gliding now in one direction, now in another; and although the flow is usually in one direction in any particular thread, no system can be discovered in the complicated movements of the whole. In the larger threads the curious spectacle often appears of two rapid currents flowing in opposite directions on opposite sides of the same thread. The currents in the thread may be seen to join currents of the peripheral layer which flow here and there, but without the regularity observed in the protoplasm of *Nitella*. The protoplasmic network also, as a whole, undergoes a slow but steady change of form, its delicate strands slowly swaying hither and thither, while the nucleus travels slowly from point to point.

Finally, we may consider an example of a form of protoplasmic movement known as *ciliary* action, which plays an important rôle in our own lives and those of lower animals and of some plants. The interior of the trachea, or windpipe, is lined by cells having the form shown in Fig. 20. At the free surface of the cell (turned towards the cavity of the trachea) the protoplasm is produced into delicate vibratory filaments having a sickle-shape when bent; these are known as *cilia* (*cilium*, an eyelash). They are so small and lash so vigorously as to be nearly or quite invisible until the movements are in some way made sluggish.

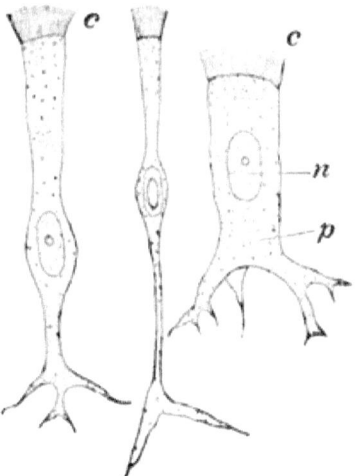

FIG. 20. (After Klein.) Three isolated ciliated cells from the interior of the windpipe of the cat. *c*, the cilia at the free end; *n*, the nucleus; *p*, the protoplasm. (Highly magnified.)

The movement is then seen to be more rapid and vigorous in one direction than in the other, all the cilia working together like the oars of a row-boat acting in concerted motion. By this action a definite current is produced in the surrounding medium (in this case the mucus of the trachea) flowing in the direction of the more vigorous movement. In the trachea this movement is upwards towards the mouth, and mucus, dust, etc., are thus removed from the lungs and windpipe. In many lower animals and plants, especially in the embryonic state, cilia are used as organs of locomotion, serving as oars to drive the organism through the water. The male reproductive germs of plants and animals are also propelled in a similar fashion.

In all these forms of vital action the protoplasm is *visibly* at work. In most cases, however, no movements of the protoplasm in cells can be detected. But it is certain from indirect evidence that protoplasm is no less active in those modes of physiological action that give no visible outward sign, as for example in an active nerve-cell or a secreting cell. This activity being molecular and chemical is beyond the reach of the microscope, but it is none the less real; and the play of these invisible molecular actions is doubtless far more tumultuous and complicated than the visible movements of the protoplasmic mass displayed in *Nitella* or in a nettle-hair. It is of the utmost importance that the student should attain to a full and vivid sense of the reality and energy of this invisible activity even in protoplasm which (as is ordinarily the case) under the closest scrutiny appears to be absolutely quiescent.

The Sources of Protoplasmic Energy. Whence comes the power required for protoplasmic action, and how is it expended? The answer to this question can be given at this point only in very general terms. It is certain that protoplasm works by means of chemical actions taking place in its own substance; and it is further certain that these actions are, broadly speaking, processes of oxidation or combustion; for in the long run all forms of protoplasmic action involve the taking up of oxygen and the liberation of carbon dioxide. Energy is therefore set free in living, active protoplasm somewhat as it is in the combustion of fuel under the boiler of a steam-engine, and in this process the protoplasm, like the coal, is gradually used up, disintegrates, and wastes away, giving off as waste matter the various chemical products of the combustion, and liberating energy as heat and mechanical work. The loss of substance is, however, continually made good (much as the coal is replenished) by the absorption of new substance in the form of food, which may consist of actual protoplasm, derived from other living beings, or of substances convertible into it. These substances are in some unexplained way converted into protoplasm and thus built into the living fabric.

To this dual process of waste ("*katabolism*") and repair ("*anabolism*") is applied the term *metabolism*, which must be considered as the most characteristic and fundamental property of living matter. It is evident from the foregoing that metabolism involves on the one hand a destructive action (*katabolism*) through which protoplasm disintegrates and energy is set free, and on the other hand a constructive action (*anabolism*) whereby new protoplasm is built up from the income of food and fresh energy is stored. It is a most remarkable fact that as far as known the constructive action resulting in the formation of new protoplasm never takes place except through the immediate agency of protoplasm already existing. In other words, there is no evidence that "spontaneous generation" or the production of living from lifeless matter without the influence of antecedent life ever takes place. Nor is there any evidence that any energy can be "generated," but rather that the vital energy of living things is only the transformed energy of their food, and that "vital force" having an origin elsewhere than in such energy does not exist.

Chemical Relations. We know nothing of the precise chemical composition of living protoplasm, because, as has been said (p. 2), living protoplasm cannot be subjected to chemical analysis without destroying its life. But the results of chemical examinations leave no doubt that the molecules of protoplasm are highly complex and are probably separated from one another by layers of water.

A. PROTEIDS. It has already been stated (p. 3) that the characteristic products of the analysis of protoplasm are the group of closely-related substances known as *proteids*. But proteids form only a small part of the total weight of any plant or animal, being always associated with quantities of other substances. Even the white of an egg, which is usually taken for a typical proteid, contains only twelve per cent of actual proteid matter, the remainder consisting chiefly of water. The following table shows the percentage of proteids and other matters in a few familiar organisms and their products:

PROXIMATE PERCENTAGE COMPOSITION OF SOME COMMON SUBSTANCES.*

Arranged according to richness in Proteids.

		Water.	Proteids.	Carbohydrates.	Fats.	Other Substances.
1	Apples	84.8	0.4	14.3	0.0	0.5
2	Indian corn, aerial portion fresh	84.3	0.9	13.7	0.5	0.6
3	Oysters, shells included	15.4	1.0	0.6	0.2	82.8
4	Turnips	91.2	1.0	6.9	0.2	0.7
5	Melons	95.2	1.1	2.5	0.6	0.6
6	Sweet potatoes	75.8	1.5	21.1	0.4	1.2
7	Crayfish, whole	10.0	1.9	0.1	0.1	87.9
8	Irish potatoes	75.5	2.0	21.3	0.2	1.0
9	Clams, round, shells included	25.3	2.1	1.3	0.1	69.2
10	Oats, aerial part fresh	81.0	2.3	15.3	0.5	0.9
11	Grass, " " "	75.0	3.0	19.9	0.8	0.3
12	Peas, " " "	81.5	3.2	13.8	0.6	0.8
13	Cow's milk	87.4	3.4	4.8	3.7	0.7
14	Flounder, whole	25.2	5.2	0.0	0.3	67.3
15	Lobster	33.0	5.4	0.2	0.5	60.9
16	Poplar and elm leaves, fresh	70.0	6.0	22.0	1.5	0.5
17	Crab, whole	34.1	7.3	0.5	0.9	56.2
18	Brook trout, whole	40.3	9.9		1.1	48.7
19	Hen's eggs, shells included	65.6	11.1	0.5	10.8	12.0
20	Mutton "chops"	41.3	12.5		29.3	16.9
21	Chicken, whole	42.2	14.3		1.1	42.4
22	Beef, heart	51.4	14.9		24.8	6.9
23	Beef, liver	60.5	20.1	3.5	5.4	1.5
24	Beefsteak, round, lean	60.0	20.7		8.1	11.2
25	Beans	13.7	23.2	55.4	2.1	3.6
26	Cheese	31.2	25.1	2.4	35.4	3.9
27	Cheese from skimmed milk	41.3	38.3	9.0	6.8	4.6

All proteids have nearly the same chemical composition and similar physical properties, however different may be the forms of protoplasm in which they occur. The analysis of protoplasm, or rather of the proteids which are its basis, teaches us really nothing of its vital properties, but serves only to show the chemical composition of the material basis by which these are manifested.

Proteids are so called from their resemblance to protein ($πρῶτος$, *first*), a hypothetical substance first described and

---

* Compiled chiefly from tables of food-composition prepared by W. O. Atwater for the Smithsonian Institution, though a few examples have been added—viz., numbers 2, 10, 11, 12, 16—from Johnson's *How Crops Grow*, N. Y., 1883.

named by Mulder. According to Hoppe-Seyler they have approximately the following percentage composition:

	C.	H.	N.	O.	S.
From	51.5	6.9	15.2	20.9	0.3
To	54.5	7.3	17.0	23.5	2.0

A small quantity of phosphorus is also very frequently present. Associated with these elements are always small quantities of various mineral substances which remain as the ash when protoplasm is burned; but the nature of their relations to the other elements is uncertain. The ash varies both in quantity and chemical composition in different animals and plants. In the white-of-egg the chief constituents of the ash are potassium chloride (KCl) and sodium chloride (NaCl), the former being much in excess. The remainder consists of phosphates, sulphates, and carbonates of sodium and potassium, with minute quantities of calcium, magnesium, and iron, and a trace of silicon. Many other mineral substances occur in association with other kinds of proteids, but always in very small proportion. These salts are in some way essential to the activity of protoplasm, as we know by familiar experience. Man, like other animals and the plants, requires certain mineral substances (e.g. common salt), but we have no knowledge of the part these play in protoplasm.

It is important to note the close chemical similarity of animal and vegetal proteids, because this is one reason for regarding vegetal and animal protoplasm as essentially similar in other respects. The following table, from Johnson after Gorup-Besanez and Ritthausen, shows the percentage composition of various proteids, and proves that the difference between vegetal and animal proteids is chemically no greater than that between different kinds of vegetal or different kinds of animal proteids:

PERCENTAGE COMPOSITION OF PROTEIDS.

	C.	H.	N.	O.	S.
Animal albumen	53.5	7.0	15.5	22.4	1.6
Vegetal "	53.4	7.1	15.6	23.0	0.9
Animal casein	53.6	7.1	15.7	22.6	1.0
Vegetal "	50.5	6.8	18.0	24.2	0.5
Animal (flesh) fibrin	54.1	7.3	16.0	21.5	1.1
Vegetal (wheat) "	54.3	7.2	16.9	20.6	1.0
Animal (blood) "	52.6	7.0	17.4	21.8	1.2

There is a corresponding likeness in the general properties and reactions of proteids. They are colloidal or non-diffusible, i.e., they will not pass through the membrane of a dialyser, or only with great difficulty; they are rarely crystalline; they rotate the plane of polarized light to the left. Though not all soluble in water, they may be dissolved by the aid of heat in strong acetic acid and in caustic alkalies, but are insoluble in cold absolute alcohol and in ether. They may be precipitated from solution by strong mineral acids, etc. Many proteids are precipitated by heat (a process which is called *coagulation*); and it is worthy of note that temperatures which produce coagulation of proteids (40°—75° C.) produce also the death of most organisms. "Amongst the organic proximate principles which enter into the composition of the tissues and organs of living beings, those belonging to the class of *proteid* or *albuminous* bodies occupy quite a peculiar place and require an exceptional treatment, for they alone are never absent from the active living cells which we recognize as the primordial structures of animal and vegetable organisms. In the plant, whilst we recognize the wide distribution of such constituents as cellulose and chlorophyl, and acknowledge their remarkable physiological importance, we at the same time are forced to admit that they occupy altogether a different position from that of the proteids of the protoplasm out of which they were evolved. We may have a plant without chlorophyl, and a vegetable cell without a cellulose wall, but our very conception of a living, functionally active, cell, whether vegetable or animal, is necessarily associated with the integrity of its protoplasm, of which the invariable organic constituents are proteids.

"In the animal, the proteids claim even more strikingly our attention than in the vegetable, in that they form a very much larger proportion of the whole organism, and of each of its tissues and organs. We may indeed say that the material substratum of the animal organism is proteid, and that it is through the agency of structures essentially proteid in nature that the chemical and mechanical processes of the body are effected. It is true that the proteids are not the only organic constituents of the tissues and organs, and that there are others, present in minute quantities, which probably are almost as widely distributed, such as for instance phosphorus-containing fatty bodies, and glycogen, yet avowedly we can (at the most) only say *probably*, and cannot, in reference to these, affirm that which we may confidently affirm of the proteids—that they are indispensable constituents of every living, active, animal tissue, and indissolubly connected with every manifestation of animal activity." (Gamgee, *Physiological Chemistry*, Chap. I.)

The molecular instability of proteids is proved by the ease with which they may be decomposed into simpler compounds; their complex constitution by the numerous compounds, themselves often highly complex, which may thus be derived or split off from them.

Amongst the other matters found in protoplasm or closely associated with it those of most frequent occurrence and greatest physiological importance are two groups of less complex substances, viz., **carbohydrates** and **fats**. These contain carbon, hydrogen, and oxygen, but no nitrogen; they do not appear to be closely related to **proteids** in chemical constitution, but they occur to some extent almost everywhere in living organisms, and in many instances are known to be of great importance, especially in nutrition. They are rich in potential energy and mobile in molecular arrangement; hence it is not strange that they figure largely in food, and are often laid by as reserve food-materials in the organism.

B. CARBOHYDRATES. These substances are so called because, besides carbon, they contain hydrogen and oxygen united in the same proportions as in water. They include starch, various kinds of sugar, cellulose, and glycogen. Starch ($C_6H_{10}O_5$) is of very frequent occurrence in plant-cells, where it appears in the form of granules embedded in the protoplasm (Fig. 9). Cellulose, having the same chemical formula as starch, but quite different in physical properties, almost invariably forms the basis of the cell-membrane in plants.

C. FATS. These are of especial importance as reserves of food-materials (*e.g.*, in adipose tissue and in seeds). They contain much less oxygen than the carbohydrates; are therefore more oxidizable, and richer in potential energy.* They commonly occur in the form of drops suspended in the protoplasm (Fig. 17), and are especially common in animal cells, though by no means confined to them.

Physical Relations. The appearance, consistency, etc., of protoplasm have already been described; but it still remains to speak of certain of its other physical properties, and especially of the manner in which its activity is conditioned by various physical agents.

Relations of Vital Action to Temperature. It is a general law that within certain limits heat accelerates, and cold diminishes, the activity of protoplasm. We know that cold tends to

---

* According to careful researches, one pound of butter contains 5654 foot-tons, and a pound of sugar 2755 foot-tons, of energy. A pound of proteid is nearly equivalent in this respect to a pound of carbohydrate.

benumb our own bodies (provided they become really chilled), and in lower animals the heart beats more slowly, the movements become sluggish or cease, breathing becomes slow and heavy,—in a word, all of the vital actions become depressed,—whenever the ordinary temperature is sufficiently lowered. If we chill the rotating protoplasm of *Chara* or *Nitella*, the vibrating cilia of ciliated cells, or an actively flowing *Amœba*, the movements become slower, and finally cease altogether.

On the other hand, moderate warmth favors protoplasmic action. Benumbed fingers become once more nimble before the warmth of the fire. In a hot room the frog's heart beats more rapidly, cilia lash more energetically, the *Amœba* flows more rapidly, and the protoplasm of *Chara* courses more swiftly. In the winter months the protoplasm of plants and of many animals is in a state of comparative inactivity. Most plants lose their leaves and stop growing; many animals bury themselves in the mud or in burrows, and pass the winter in a deep sleep (*hibernation*), during which the vital fires burn low and seem well-nigh extinguished. The warmth of spring re-establishes the activity of the protoplasm, and in consequence animals awake from their sleep and plants put forth their leaves.

But this law is true only within certain limits. Extreme heat and cold are alike inimical to life, and as the temperature approaches these extremes all forms of vital action gradually or suddenly cease. The limits are so variable that it is not at present possible to formulate any exact law which shall include all known cases. For instance, many organisms are killed at the freezing-point of water ($0°$ C.); but certain forms of life have withstood a temperature of $-87°$ C. ($-123°$ F.), and recent experiments show that frogs and rabbits may be chilled to an unexpected degree without fatal results.

The upper limit is also inconstant, though less so than the lower. Most organisms are destroyed at the temperature of boiling water ($100°$C.), but the spores of bacteria have been exposed to a much higher temperature without destruction ($120°$–$125°$ C.). As a rule, protoplasm is killed by a temperature varying from $40°$ to $50°$ C., the immediate cause of death being apparently due to a sudden coagulation (p. 36) of certain substances in the protoplasm. Thus, if a brainless frog be gradually heated, death ensues at about 40° C., and the body becomes stiff and rigid (*rigor caloris*) from the coagulation of the muscle-substance. The lower forms of animal life agree well with plants in their "fatal temperatures," which in many cases lie between 40° and 50° C.

Lastly, it appears to be true that there is a certain most favorable or *optimum* temperature for the protoplasm of each species of plant and animal, this optimum differing considerably in different species. Probably the highest limit occurs among the birds, where the uniform temperature of the body may be as high as 40° C. The lowest occurs among the marine plants and animals of the Arctic seas, or of great depths, where the temperature seldom rises more than a degree or two above the freezing-point. Between these limits there appears to be great variation, but 35° C. may perhaps be taken as the average optimum.

Moisture. Protoplasm always contains a large amount of water, of which indeed the lifeless portion of living things chiefly consists. (See table on p. 34.) All plants and animals are believed to be killed by complete drying, though some of the simpler forms resist partial drying for a long time, becoming quiescent and reviving again when moistened, sometimes even after the lapse of years. Hence water appears to be an essential constituent of protoplasm, although, as in the case of mineral matters, we do not know the nature of its connection with the other elements or compounds present.

Electricity. It has been shown that many forms of vital action are accompanied by electrical disturbances in the protoplasm. It is therefore not surprising that the application of electricity to living protoplasm should have a marked effect on its actions. If the stimulus be very slight, protoplasmic movements are favored. Colorless blood-corpuscles creep more actively, and ciliary action increases in vigor. Stronger shocks cause a spasmodic contraction of the protoplasm (*tetanus*), from which it may or may not be able to recover, according to the strength of the shock.

Poisons. Towards certain agents protoplasm is indifferent or seemingly so, but towards others it behaves in a very remarkable manner. The matters known as poisons modify or destroy its activity, as is well known from the familiar effects of arsenic, opium, etc. Disease may also interfere with its normal activity; but the consideration of these phases of the subject belongs to the more exclusively medical sciences, such as toxicology and pathology.

Other Physical Agents. The more highly specialized forms of protoplasm are affected by a great variety of physical agents, such as light, sound, pressure, etc., and upon this susceptibility depend many of the higher manifestations of life. For instance, waves of light or of sound, acting upon special protoplasmic structures in the eye and ear, call forth actions which ultimately result in the sensations of sight and hearing. Similar considerations apply to the senses of smell, taste, and touch; but the discussion of all these special modes of protoplasmic action must be deferred. Enough has been said to show that living organisms (that is, the protoplasm which is their essential part) are able to respond to many influences proceeding from the world in which they live. Upon this property depend the intimate relations between the organism and its environment, and the power of adaptability to the environment which is one of the most marvellous and characteristic properties of living things.

Non-diffusibility. Living protoplasm, like most of the various proteid matters which it yields (p. 36), is *indiffusible*. It will be seen eventually that osmotic processes play a leading rôle in the lives of plants and animals, since they are in large part the means by which nutriment is conveyed to the living substance. In view of this fact, the non-diffusibility of protoplasm as well as of ordinary proteids is a fact of much significance.

Vegetal and Animal Protoplasm. The protoplasm of plants is essentially identical with that of animals in chemical and physical relations, and manifests the same fundamental vital properties. But it would manifestly be absurd to suppose this identity absolute, for if it were so, plants and animals would also be identical; and furthermore, the protoplasm of every species of plant and animal must differ more or less from the protoplasm of every other species. What is meant is that the differences between the many kinds of protoplasm are far less important than the fundamental resemblances which underlie them.

CHAPTER IV.

THE BIOLOGY OF AN ANIMAL.

The Common Earthworm.
(*Lumbricus terrestris*, Linnæus.)

We now advance to a more precise examination of the living body considered as an individual. It is a familiar fact that living things fall into two great groups, known as plants and animals. We shall therefore examine a representative of each of these grand divisions of the living world, and inquire how they resemble each other and how they differ. Any higher animal would serve as a type, but the common earthworm is a peculiarly favorable object of study, because of the simplicity of its structure, the clearness of its relation to other animals standing above and below it in the scale of organization, and the ease with which it may be procured and dissected. Earthworms, of which there are many kinds, are found in all parts of the world, extending even to isolated oceanic islands. In the United States there are several species, of which the most common are *L. communis* (*Allolobophora mucosa*, Eisen), *L. terrestris*, and *L. fœtidus* (*Allolopobhora fœtida*, Eisen). The first two of these are found in the soil of gardens, etc., *L. terrestris* being the larger and stouter species and readily distinguishable by the flattened shape of the posterior region. *L. fœtidus*, a smaller red species, transversely striped, and having a characteristic odor, occurs in and about compost-heaps.

Mode of Life, etc. Earthworms live in the earth, burrowing through the soil at a depth varying from a few inches to several feet. Here they pass the daytime, crawling out at night or after a shower. The burrows proceed at first straight downwards, and then wind about irregularly, sometimes reaching a depth of six or eight feet. The earthworm is a nocturnal animal, and during the day lies quiet in its burrow near the surface, extended at full length, head uppermost. At night it becomes very active, and, thrusting the fore end of the body far out, explores the vicinity in all directions, though still clinging fast, as a rule, to the mouth of the burrow by the hinder end. In this way the worm is able to forage, seizing leaves, pebbles, and other small objects, and dragging them into the burrow. Some of these are devoured; the remainder (including the pebbles, etc.) are used to line the upper part of the burrow, and to plug up its opening when the worm retires for the day. Besides bits of leaves and animal matter, earthworms swallow large quantities of earth, which is passed slowly through the alimentary canal, so that any nutritious substances contained in it may be digested and absorbed. This earth is generally swallowed at a considerable distance below the surface of the ground, and is finally voided at the surface near the opening of the burrow. In this way arise the small piles of earth ("*castings*" or *fæces*) which every one has seen, especially in the morning, wherever earthworms abound. Very large quantities of earth are thus brought to the surface by earthworms—in some cases, according to Darwin's estimates, more than eighteen tons per acre in a single year. In fact, most soils are continually being worked over by worms; and Darwin has shown that these humble creatures, in the course of centuries, have helped to bury huge rocks and the ruins of ancient buildings.*

The earthworm has no ears, eyes, or any other well-marked organs of special sense. Nevertheless—and this is a point of great physiological interest—the fore end of the body is sensitive to light; for if a strong light be suddenly flashed upon this part of the worm as it lies stretched forth, it will often "dash like a rabbit into its burrow." The animal has a keen sense of touch, as may be proved by tickling it; and its sense of taste must be well developed, since the worm is somewhat fastidious in its choice of food. Earthworms appear to be quite deaf, but possess a distinct, though feeble, sense of smell.

---

* Darwin, *Vegetable Mould and Earthworms*. Appleton, N. Y., 1882. See also White's *Natural History of Selborne*, Index, references to "Earthworms."

General Morphology.

Attention will first be directed to certain features of the BODY seemingly of little importance, but really full of meaning when compared with like features in other animals higher or lower in the scale of organization.

Antero-posterior Differentiation. The body (Fig. 21) has an elongated cylindrical form, tapering to a blunt point at one end, obtusely rounded and flattened at the other. As a rule, the pointed end moves forwards in locomotion, and the mouth opens near it. For these and other reasons the pointed end might be called the head-end, and the other the tail-end. But the worm has really neither head nor tail, and hence the two ends may better be distinguished as the *fore end* and the *hinder end*, or still better as *anterior* and *posterior*. And in scientific language the fact that the worm has anterior and posterior ends which differ from each other is stated by saying that it shows *antero-posterior differentiation*. This simple fact acquires great importance in the light of comparative biology; for it may be shown that the antero-posterior differentiation of the earth-worm, insignificant as it seems, is only the begining of a series of important modifications extending upwards through more and more complex stages to culminate in man himself.

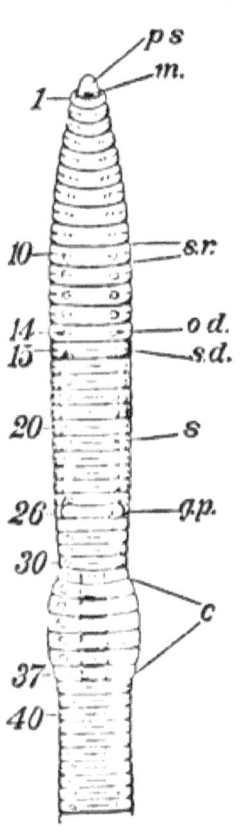

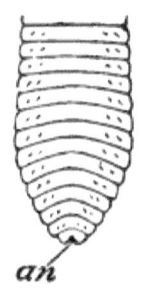

FIG. 21.—Enlarged view of the anterior and posterior parts of the body of an earthworm as seen from the ventral aspect. *an*, anus; *c*, clitellum; *g.p.*, glandular prominences on the 26th somite; *m*, mouth; *o.d.*, external openings of the oviducts; *p.s.*, prostomium; *s*, setæ; *s.r.*, openings of the seminal receptacles; *s.d.*, external openings of the sperm-ducts. The form of the body varies greatly in life according to the state of expansion. The specimen here shown is from an alcoholic preparation. (Slightly enlarged.)

Dorso-ventral Differentiation. In living or well-preserved specimens, the body is not perfectly cylindrical, but is somewhat flattened, particularly near the posterior end, and has a slightly prismatic four-sided form. One of the flattened sides, slightly darker in color than the other, is habitually turned upwards, and is therefore called the back, the opposite or lower side, commonly turned downwards, being the belly. For the sake of accuracy, however, biologists are wont to speak of the *dorsal aspect* (back) and *ventral aspect* (belly) of the body; and the fact that an animal has a back and belly differing from each other in structure or function, or both, as in the earthworm, is expressed by saying that the body exhibits *dorso-ventral differentiation*. This, like antero-posterior differentiation, is very feebly expressed in the external features, though clearly marked in the arrangement of the internal parts of the earthworm. In higher animals it becomes one of the most conspicuous features of the body.

Bilateral Symmetry. When the body is placed in the natural position, with the ventral aspect downwards, a vertical plane passing longitudinally through the middle will divide it into exactly similar right and left halves. This similarity is called two-sided likeness, or *bilateral symmetry*. Though not very obvious externally, this symmetry characterizes the arrangement of all the internal parts; and it may be gradually traced upwards in higher animals, until it becomes as striking and perfect as in the human body.

Thus a very superficial examination reveals in the earthworm two fundamental laws of organization, viz., *differentiation* or the law of difference, and *symmetry* or the law of likeness. And these laws are of interest for the reason among many others that earthworms, like other organisms, have as a race had a history, have *come to be* by a gradual process (cf. p. 99). And biology must strive to answer the questions *how* and *why* certain parts have become symmetrical and others differentiated. Without entering into a full discussion of the question at this point, it may be said that the main cause of symmetry or differentiation has probably been likeness or unlikeness of function, or of relation to the environment. Earthworms show antero-posterior and dorso-ventral differentiation, because the anterior and posterior extremities, or the dorsal and ventral aspects, have been differently used and exposed to different conditions of environment. And on the other hand the organism is bilaterally symmetrical, because the two sides have been similarly used and have been exposed to like conditions of environment.

Metamerism. Another general feature of the earthworm is of great importance in view of the conditions existing in other animals, including the higher forms. The body is marked off by transverse grooves into a series of similar parts like the joints of a bamboo fishing-rod, or like the joints of fingers (Fig. 21). These parts are called *metameres*, or more often *somites*, and the body is consequently said to have a *metameric* structure, or to exhibit *metamerism*. From the outside, the somites appear to be produced simply by regular folds in the skin, like the wrinkles between the joints of our fingers. But as the wrinkles of the fingers are only the external expression of a more fundamental jointed structure within, so the external folds separating the somites, represent an internal division into successive parts, which affects all the organs of the body, and is a result of some of the most important phenomena of development.

The explanation of metamerism or "*serial symmetry*" is one of the most difficult problems of morphology. But it will be seen farther on that metamerism, so clearly and simply expressed in the earthworm, can be traced upward in ever-increasing complexity to the highest forms of life, and suggests some of the most interesting and fundamental problems with which biology—and especially morphology—has to deal. Indeed, the comparative study of the anatomy of most higher animals consists very largely in tracing out the manifold transformations of their complicated somites, which under many disguises can be recognized as fundamentally like the simpler somites of the earthworm.

Modifications of the Somites. The somites differ considerably in different parts of the body. The extreme anterior end is formed by a smoothly-rounded knob called the *prostomium*, which is shown by its mode of development not to be a true somite. It forms a kind of overhanging upper lip to the *mouth*, which lies just behind it on the ventral aspect. Behind the mouth is the first somite, in the form of a ring,* interrupted above by a backward prolongation of the prostomium.

---

* In numbering the somites the prostomium must never be reckoned, the first somite being *behind the mouth*.

The somites from the 1st to the 27th are rather broad, and gradually increase in size. A variable number of the somites lying between the 7th and 19th are often swollen on the ventral side, forming the so-called *capsulogenous glands*. Between the 28th and 35th (the number and position varying slightly in different specimens) the somites are swollen above and on the sides, and the folds between them are scarcely defined except on the ventral aspect. Taken together, they form a broad, conspicuous girdle called the *clitellum* (Fig. 21, *c*), whose function is to secrete the capsule in which the eggs are laid, and also a nutritive milk-like fluid for the use of the developing embryos. (The clitellum is not present in immature specimens.) Behind the clitellum the somites are narrower, somewhat four-sided in cross-section, and flattened from above downwards. This flattening sometimes becomes very conspicuous towards the posterior end. Towards the very last they decrease in size rather abruptly, and they end in the *anal* somite, which is perforated by a vertical slit, the *anus* (Fig. 21, *an*). All the somites are perforated by small openings leading into the interior of the body, and forming the outlets of numerous organs; the position of these openings will be described in treating of the organs. Each somite, excepting the anterior two or three and the last, gives insertion to four groups of short and minute bristles or *setæ*, which are arranged in four longitudinal rows along the body. Two of these rows run along the ventral aspect, two are more upon the sides. The setæ extend outwards from the interior of the body, where they are supplied with small muscles by which they can be turned somewhat either forwards or backwards, and can also be protruded or withdrawn (Fig. 22). The setæ are of great use in locomotion. When pointed backwards they support the worm as it crawls forwards; when they are turned forwards the worm can creep backwards. They are of interest, therefore, as representing an extremely simple and primitive limb-like organ.

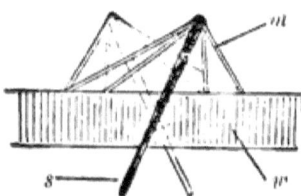

FIG. 22.—Diagram to illustrate the action of the setæ. The dotted outline represents the position of the seta and its muscles when bent in the opposite direction. *m*, muscles; *s*, seta; *w*, body-wall.

GENERAL PLAN OF THE BODY. 47

Plan of the Body. The body of the earthworm (Fig. 23), like that of all higher animals, consists of two tubes, one (*al*) within the other and separated from it by a considerable space or cavity (*cœ*). The inner tube is the *alimentary canal*, opening in front by the *mouth* and behind by the *anus*; the outer tube is the body-wall, and its cavity is the *body-cavity* or *cœlom*.

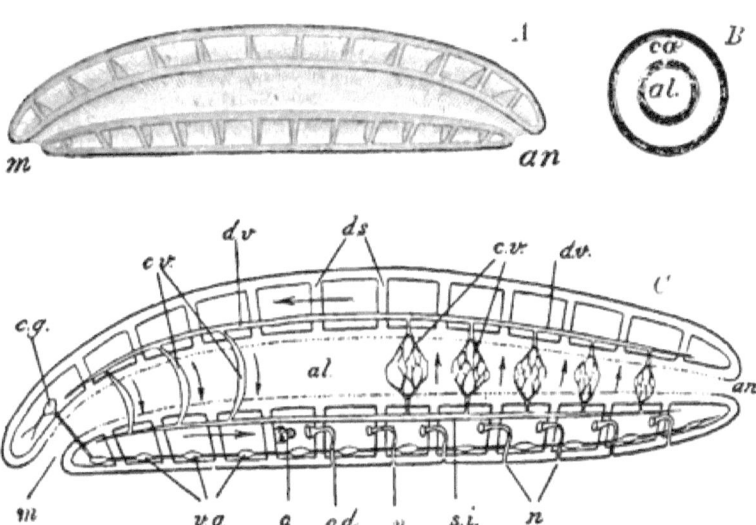

Fig. 23.—*A*, diagram of the earthworm as seen in a longitudinal section of the body, showing the two tubes, the cœlom, and the dissepiments. *B*, diagram of cross-section: *al*, alimentary tube; *an*, anus; *cœ*, cœlom; *m*, mouth. *C*, diagram showing the arrangement of some of the principal organs: *m*, mouth; *an*, anus; *al*, alimentary canal; *ds*, dissepiments; *d.v.*, dorsal blood-vessel; *v*, ventral or sub-intestinal vessel; *c.v.*, circular vessels; *n*, nephridia or excretory organs; *c.g.*, cerebral ganglia; *v.g.*, ventral chain of ganglia; *o.d.*, oviduct; *o.d.*, ovary. The arrows indicate the course of the circulation of the blood.

The cœlom is not, however, a free continuous space extending from end to end, but is divided transversely by a series of thin muscular partitions, the *dissepiments*, into a series of nearly closed chambers traversed by the alimentary canal. Each compartment corresponds to one somite, the dissepiments being opposite the external furrows mentioned on p. 45. All the organs of the body are originally developed from the walls of these chambers, and some of them (e.g., the organs of excretion) project into the cavities of the chambers, that is into the cœlom.

In the median dorsal line of each somite (excepting the first two or three) is a minute pore (the *dorsal pore*) which perforates the body-wall and thus places the cœlom in connection with the exterior.* Other pores that pass through the body-wall into the cavities of various organs will be described further on.

Organs of the Animal Body. Systems of Organs. The body of the earthworm consists essentially of protoplasm, and in order that so large a mass of living matter may continue to exist and carry on the ordinary life of an earthworm it must be able to obtain a sufficient supply of food; to digest and absorb it, and distribute it to all parts of the body; to build up new protoplasm and remove waste. It must be sensitive to external and internal influences; capable of motion and locomotion. Above all, each part must act with reference to, and in harmony with, every other part, so that the organism may not be merely an aggregate of organs, but one body acting as a unit or a whole.

These *functions* are fulfilled by the ORGANS, respectively, OF ALIMENTATION, DIGESTION, ABSORPTION, CIRCULATION, EXCRETION, SENSATION, MOTION, and COÖRDINATION. All of these minister to the welfare of the individual. The REPRODUCTIVE function, on the other hand, and its corresponding organs, serve to perpetuate the species, thus ministering rather to the race than to the individual.

Sets of organs devoted to the same function constitute *systems;* as the *alimentary system*, the *circulatory system*, etc. Those which are more immediately concerned with the income and outgo of matter—namely, the alimentary, digestive, absorptive, circulatory, and excretory systems—are sometimes called the *vegetative systems* or *systems of nutrition;* while those which have to do more immediately with the relation of the body to its environment, rather than the individual itself, are called *systems of relation*. Examples of the latter are the systems of organs of support, motion (including locomotion), sensation, and coördination; and even the reproductive system, as relating chiefly to other individuals, finds a place here.

---

* If living worms be irritated they will often extrude a milky fluid from these pores, but the use of the latter is not well understood.

A. Systems of Nutritive Organs: their Special Morphology and Physiology. (For B see p. 62.)

Alimentary System (Organs of Alimentation). Earth-worms feed mainly upon leaves or decaying vegetable matter, but will also eagerly devour meat, fat, and other animal substances. They also swallow large quantities of earth from which they extract not only any organic materials that it may contain, but probably also moisture and a small amount of various salts. The most essential and characteristic part of their food is derived from vegetal or animal matter in the form of various organic compounds, of which the most important are *proteids* (protoplasm, albumen, etc.), *carbohydrates* (starch, cellulose), and *fats*. These materials are used by the animal in the manufacture of new protoplasm to take the place of that which has been used up. It is, however, impossible for the animal to build these materials directly into the substance of its own body. They must first undergo certain preparatory chemical changes known collectively as *digestion*; and only after the completion of this process can all the food be absorbed into the circulation. For this purpose the food is taken not into the body proper, but into a kind of tubular chemical laboratory called the *alimentary canal* through which it slowly passes, being subjected meanwhile to the action of certain chemical substances, or reagents, known as *digestive ferments*. These substances, which are dissolved in a watery liquid to form the *digestive fluid*, are secreted by the walls of the alimentary tube. Through their action the solid portions are liquefied and the food is rendered capable of absorption into the proper body.

The alimentary canal is divisible into several differently constructed portions playing different parts in the process of alimentation. Going backwards from the mouth these are as follows:

1. The *pharynx* (Fig. 24, *ph*), an elongated barrel-shaped pouch extending to about the 6th somite. Its walls are thick and muscular, and from their cœlomic surface numerous small muscles radiate on every side to the body-wall. When these muscles contract, the cavity of the pharynx is expanded; and if the mouth has been previously applied to any solid object, such as a leaf or pebble, the pharynx acts upon it like a suction-pump.

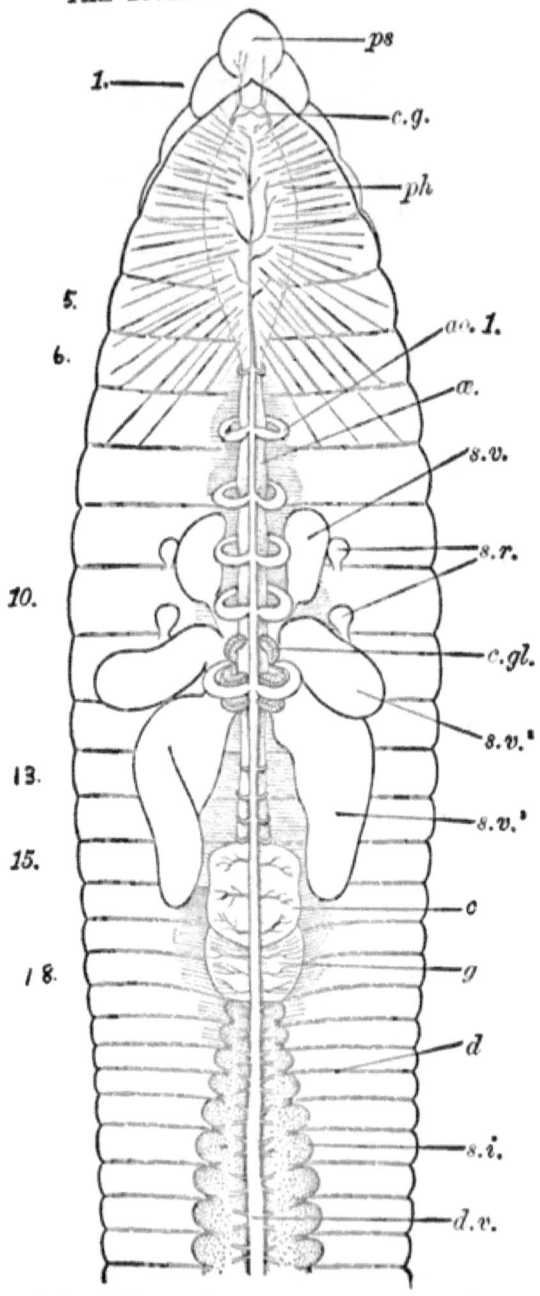

Fig. 24.—Dorsal view of the anterior part of the body of *Lumbricus*, as it appears when laid open along the dorsal aspect. *ao*, aortic arch; *c*, crop; *c.g*, cerebral ganglia; *c.gl*, calciferous glands; *d*, dissepiment; *d.v*, dorsal vessel; *g*, gizzard; *œ*, œsophagus; *ph*, pharynx; *ps*, prostomium; *s.i*, stomach-intestine, showing the lateral pouches; *s.r*, seminal receptacles; $s.v.^1$, $s.v.^2$, $s.v.^3$, the three pairs of lateral seminal vesicles.

In this way the animal lays hold of the various objects, nutritious and otherwise, which it devours or draws into its burrow.

Embedded in the muscular walls of the pharynx are a number of small "salivary" glands of whose function nothing is definitely known, though they doubtless pour a digestive fluid into the pharyngeal cavity.

2. The *œsophagus* (*œ*), a slender, thin-walled tube extending from the 6th to the 15th somite. Through this the food is swallowed, being driven slowly along by wavelike (*peristaltic*) contractions (p. 55). In the region of the 11th and 12th somites are three pairs of small pouches opening at the sides of the œsophagus. These are the *calciferous glands* (*c.gl.*). They contain solid masses of calcium carbonate, and Darwin conjectures that their use is partly to aid digestion by neutralizing the acids generated during the digestion of leaves, and perhaps partly to serve as an outlet for the excess of lime in the body, especially when worms live in calcareous soil.

3. The *crop* (*c*), about the 16th somite; a thin-walled, sac-like dilatation of the alimentary canal, which serves as a reservoir to receive the swallowed food.

4. The *gizzard* (*g*), about the 17th somite; a cylindrical, firm and muscular portion, lined by a horny membrane. In this the food is rolled about, squeezed and ground to prepare it for digestion in the following portion, viz. :

5. *The stomach-intestine* (*s.i.*), which corresponds physiologically to both the stomach and intestine of higher animals. This is a straight thin-walled tube, extending from the gizzard to the anus, without convolutions, not differentiated into stomach and intestine, and devoid of distinct glandular appendages such as the liver or pancreas existing in the higher animals. The digestive fluid is secreted by the walls of the alimentary canal itself, the surface of which is much increased by the presence of lateral pouches or diverticula, one on either side in each somite. In front these are large and conspicuous, but behind they gradually diminish in size until scarcely perceptible.

The inner surface of the stomach-intestine is further increased by a deep inward fold, called the *typhlosole*, running longitudinally along the dorsal median line. The typhlosole is not visible on the exterior, but is seen by opening the stomach-intestine from the side or below, or upon making a cross-section. It is richly supplied with *blood-vessels* that pass down into its cavity from the dorsal vessel (Fig. 39), and its main function is probably to increase the surface for the absorption of food (cf. the "spiral valve" in the intestine of sharks.)

The outer surface of the stomach-intestine is covered with pigmented, yellowish-brown "chloragogue-cells." These were formerly supposed to be concerned with the secretion of the digestive fluid, and hence are often called "hepatic cells." This, however, is probably an erroneous interpretation, and they are now believed to be concerned with the process of excretion (p. 61).

Digestion. Digestion begins even before the food is taken into the alimentary canal; before being swallowed, the leaves, etc., are moistened by digestive fluid poured out from the mouths of the worms. The main action, however, doubtless goes on in the anterior part of the stomach-intestine and diminishes as the food passes backward. It has been proved by experiment that the digestive fluid acts on at least two of the three principal varieties of organic food-stuffs, viz., on proteids and on starch (carbohydrate), and in so far resembles the pancreatic fluid of higher animals, which it further resembles in having an alkaline reaction. Analogy leads us to believe that the digestive fluid has some action also on fats; but this has not been proved.

Krukenberg and Frédericq have shown that the digestive fluid of the earthworm contains at least three ferments; and according to the former author these occur only in the stomach-intestine. They are as follows:

1. *Peptic ferment*, which has the property in an acid medium of converting proteids into soluble and diffusible *peptones;* this is therefore analogous to the pepsin of the gastric juice in higher forms.

2. *Tryptic ferment*, having a similar action on proteids, but only in an alkaline medium—hence analogous to the *trypsin* of pancreatic juice.

3. *Diastatic ferment*, which converts starch into glucose (grape-sugar) in an alkaline medium—hence analogous to the ptyalin of saliva and the amylolytic ferment of pancreatic juice.

Absorption. The ferments of the digestive fluid convert the solid proteids into soluble and diffusible peptones, the starchy matters into sugar (glucose). These products dissolve in the liquids present and are then gradually absorbed by the walls of the intestine as the food passes along the alimentary canal. The precise mechanism of absorption is not yet thoroughly understood, but it is probable that much of the nutriment passes by diffusion (osmosis) into the walls of the stomach-intestine and thence into the blood for distribution to all parts of the body. The refuse remaining in the alimentary canal (and which has never been a part of the body proper) is finally voided through the anus as *castings* or *fæces*. This process of "defæcation" must not be confounded with that of *excretion*, which will be described later.

Circulatory System. The food, having been absorbed, is distributed throughout the body by two devices.

1. *Cœlomic Circulation.* The cavity of the cœlom is filled with a colorless fluid ("cœlomic fluid") which must be regarded as a kind of lymph or blood. By the contractions of the body-wall, as the worm crawls about, the cœlomic fluid is driven back and forth through all parts of the cœlom, through irregular openings in the dissepiments. As the digested food is absorbed from the stomach-intestine a considerable part of it is believed to pass into the cœlomic fluid, and is thus conveyed directly to the organs which this fluid bathes. The cœlomic fluid is composed of two constituents, viz., a colorless fluid called the *plasma*, and colorless isolated cells or *corpuscles* which float in the plasma, and are remarkable for the fact that they undergo constant though slow changes of form. In fact they closely resemble certain kinds of *Amœba*, and we should certainly consider them to be such if we found them occurring free in stagnant water. We know, however, that they live only in the *plasma*, and have a common origin with the other cells of the body; hence we must regard them not as individual animals, but as constituent cells of the earthworm. The cœlomic fluid is in fact a kind of *tissue* consisting of isolated colorless cells floating in a fluid intercellular substance. These free floating cells are probably the scavengers (*phagocytes*) of the body, devouring and destroying waste matters. Some

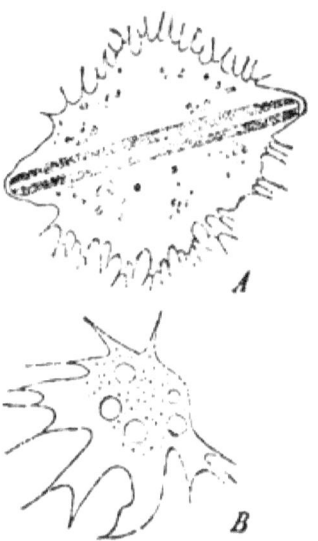

FIG. 25.— Phagocytes, from the cœlomic fluid of the earthworm. *A*, agglomeration of phagocytes, surrounding a foreign body; *B*, single phagocyte, with vacuoles. (After Metschnikoff.)

suppose that they also attack invading parasites such as bacteria.

2. *Vascular Circulation.* Besides the cœlomic circulation there is another and more complicated circulatory apparatus consisting of branching tubes, the *blood-vessels*, which form a complicated system ramifying throughout the body. Through these tubes is driven a red fluid analogous to the red blood of higher animals, and like it consisting of *plasma* and *corpuscles*, the latter being flattened and somewhat spindle-shaped. The red color is due to a substance, *hæmoglobin*, dissolved in the plasma and not (as in higher forms) contained in the corpuscles, which are colorless.

The earthworm is not provided with a special pumping-organ or heart for the propulsion of the blood, such as we find in higher animals. In place of this certain of the larger blood-vessels (viz., the "dorsal vessel" and the "aortic arches") have muscular contractile walls, which propel the blood in a constant direction by wave-like contractions that run along the vessel from one end to the other ("peristaltic" contractions, cf. p. 51) at regular intervals and thus give rise to a "pulse." The contractile vessels give off other non-contractile trunks which divide and subdivide into tubes of extremely small calibre and having very thin walls. The ultimate branches, known as *capillaries*, permeate nearly all the organs and tissues, in which they form a close network. The stream of blood after passing through the capillaries is gathered into successively larger vessels which after a longer or shorter course finally empty into the original contractile trunks and complete the circuit. Thus the vascular system is a closed system of tubes, and there is reason to believe that the blood follows a perfectly definite course, though this is not yet precisely determined.*

We may now consider the arrangement of the principal trunks. The largest of them, which is also the most important of the contractile vessels, is:

a. The *dorsal vessel* (Fig. 24, *d. v.*), a long muscular tube lying upon the upper side of the alimentary canal. In the living worm it may be distinctly seen through the semi-transparent

---

* It should be noted that in the absence of a heart it is difficult to distinguish between "arteries" and "veins." We may more conveniently distinguish "afferent vessels," carrying blood towards the capillaries, and "efferent vessels," carrying blood away from them.

skin as a dark-red band, which is tolerably straight when the worm is extended, but is made zigzag by contraction of the body. If it be closely observed, a sort of wavelike contraction is often seen running from behind forwards. This may be very clearly observed in a worm stupefied by chloroform, especially if it has been laid open along the dorsal side. The dorsal vessel then appears as a deep-red, somewhat twisted, tube running along the upper side of the alimentary canal. Wavelike contractions continually start from its hinder end and run rapidly forwards, one after another, to the anterior end, where the dorsal vessel finally breaks up on the pharynx into a large number of branches (Fig. 24).

The result of these orderly progressive contractions is that the fluid within the tube is pushed forwards—very much as the fluid in a rubber tube is forced along when the tube is stripped through the fingers. It is still better illustrated by the action of the fingers in the operation of *milking*. This action of the vessels is a typical example of *peristaltic contraction*.

b. Sub-intestinal vessel. This is a straight vessel which runs along the middle line on the *lower* side of the alimentary canal, parallel to the one just described. It returns to the hinder part of the body the fluid which has been carried forwards by the dorsal vessel. On the pharynx it breaks up into many branches, which receive the fluid from corresponding branches of the dorsal vessel.

c. Circular or commissural vessels, metamerically repeated trunks which run from the dorsal vessel downwards around the alimentary canal and ultimately connect with the ventral vessel. They are of several kinds, of which the most important are as follows:

1. The *aortic arches* or *circumœsophageal vessels*, often known as "hearts," since like the dorsal vessel they are contractile and with the latter furnish the entire propulsive force for the circulation. These are five pairs of large vessels encircling the œsophagus in somites 7 to 11 inclusive. These vessels pass directly from the dorsal to the ventral vessel, giving off no branches. During life they perform powerful peristaltic contractions, receiving blood from the dorsal vessel and pumping it into the sub-intestinal or ventral.

2. *Dorso-intestinal vessels*, passing from the dorsal vessel into the wall of the gut in the region of the stomach-intestine. Of these vessels there are two or three pairs in each somite. They are thickly covered (like the dorsal vessel in this region) with pigmented "chloragogue-cells," so that their red color is usually not apparent. Unlike the aortic arches these vessels break up on the wall of the intestine into capillaries which are continuous with branches from the ventral vessel.

3. *Dorso-tegumentary vessels*, passing from the dorsal vessel along the dissepiment into the body-wall on each side. These are small vessels that pass directly around the body to connect with a longitudinal trunk ("sub-neural") lying below the ventral nerve-cord (see below), and giving off branches to the body-wall, dissepiments, and nephridia.

Course of the Blood. The precise course of the blood in *Lumbricus* is still in dispute, though its more general features are known. It is certain that the bulk of the blood passes forward in the dorsal vessel, downward around the gut through the aortic arches into the ventral vessel, and thence backwards towards the posterior region. Its path thence into the dorsal vessel is doubtful. The most probable view is that the blood proceeds from the ventral vessel through ventro-intestinal vessels to the capillaries of the intestine and thence to the dorsal vessel through the dorso-intestinal vessels. It is possible, however, that the return path is through the dorso-tegumentary vessels and that the dorso intestinal carry blood *from* the dorsal vessel *to* the intestine.

In the foregoing account only the more obvious features of the blood-vessels have been mentioned, and many important details have been passed over. The circular vessels of the stomach-intestine can be followed for only a short distance out from the dorsal vessel, where they seem to break up into a large number of small parallel vessels lying close together and running around to the lower side. The efferent vessels do not directly join the sub-intestinal, but empty into a sinus or vessel which runs parallel to tne latter, closely imbedded in the wall of the stomach-intestine. The sub-intestinal vessel proper is quite separate from the stomach-intestine, and communicates by short branches (usually two in each somite) with the vessel lying above it. This may be clearly seen in the region of the gizzard. On this there is a variable number of small lateral vessels, which break up partly into a branching network, and are partly resolved into extremely fine parallel vessels surrounding the organ. On the crop are three or four pairs of lateral branches from the dorsal vessel which branch out into a fine network, but do not break up into parallel vessels as on the gizzard. In the two somites (13th and 14th) in front of the crop there are usually two pairs of vessels running around the œsophagus. In the 11th and 12th somites a small branch is given off to each calciferous gland. The most anterior pair of circular vessels are in the 6th somite, and are very small. In front of this the dorsal vessel breaks up into the pharyngeal network. In front of the 11th somite there are three sub-intestinal vessels. The two additional vessels lie, one on either side of the primary one and break up into branches at the sides of the pharynx. The aortic arches empty into the middle vessel, and at the point of junction there is a communication with the lateral vessel of the corresponding side.

Besides the dorsal and sub-intestinal vessels there are three other minor longitudinal trunks (Fig. 26). Two of these are very small, and lie on

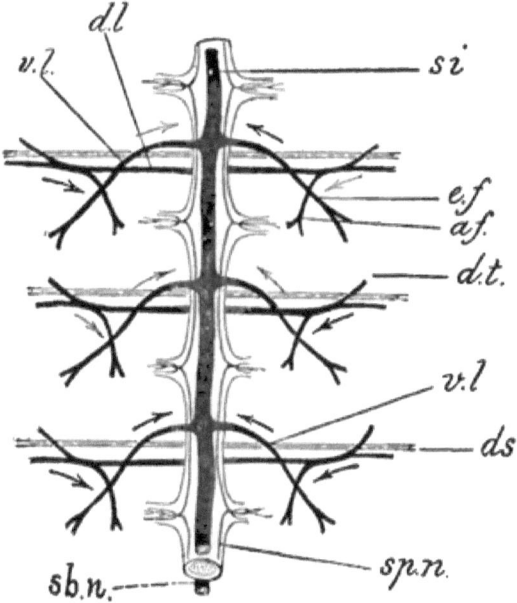

FIG. 26.—Dorsal view of part of the ventral nerve-cord, showing the arrangement of the vessels of the ventral region. *ds*, dissepiment; *si*, sub-intestinal or ventral blood-vessel; *sb.n.*, sub-neural; *sp.n.*, supra-neural. The sub-intestinal receives on either side the ventro-laterals (*v.l*) from the nephridia, of which it forms the efferent vessel (*e.f*). The sub-neural is joined on each side by a continuation of the dorso-tegumentary (*d.t.*); *af*, afferent branch to the nephridium (cf. Fig. 27).

either side above the nerve-cord (p. 66), sending fine branches out from each ganglion along the lateral nerves. These are the *supra-neural* trunks (*s.n.*). The third longitudinal vessel (*sub-neural*) lies below the nerve-cord. (See Fig. 26.) It receives on each side the termination of the dorso-tegumentary vessel (*d.t.*, Fig. 26) which in its course is connected with the capillary networks of the body-wall and the dissepiment, and gives off a large branch to the nephridium (cf. Fig. 27).

Besides the lateral vessels from the sub-neural and supra-neural a pair of "ventro-lateral" (v.l., Figs. 26 and 27) are given off in each somite from the sub-intestinal to the nephridium, probably receiving from it the blood which originally entered through a branch of the dorso-tegumentary.

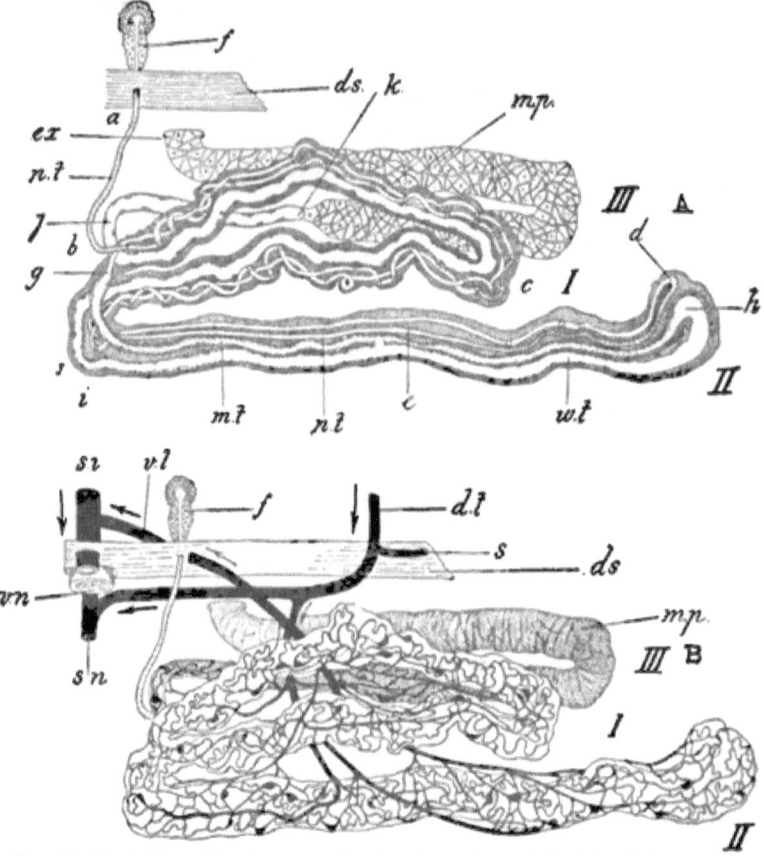

FIG. 27.—Nephridia of *Lumbricus*. A showing the regions of the tube, B the vascular supply. I, II, III, the three principal loops.
A. f, funnel; n.t, the "narrow tube"; m.t, middle tube; w.t, wide tube; m.p, muscular tube or end-vesicle, ds, dissepiment. The narrow tube extends from a to g and is ciliated between a and b, at c, and from d to e. The middle (ciliated) tube extends from g to h, the wide tube from h to k, where it opens into the muscular part; ex, external opening.
B. Letters as before; d.t, dorso-tegumentary vessel, bringing blood from the dorsal vessel, receiving at s a branch from the body-wall, sending an afferent branch to the nephridium, and finally joining the sub-neural (s.n); v.l, ventro-lateral vessel carrying the blood from the nephridium to the sub-intestinal or ventral vessel (s.i), v.n, ventral nerve-cord. (After Benham; the direction of the blood-currents according to Bourne.)

Excretory System. It is the office of the excretory system to remove from the body proper the waste matters ultimately re-

ORGANS OF EXCRETION. NEPHRIDIA.

sulting from the breaking down of living tissue. This does not mean the passing away of the refuse of digestion through the anus (defaecation, p. 53), for such matters have never been absorbed and therefore have never really been within the body proper. Excretion means the removal from the body of matter which has really formed a part of its substance, but has been used up and is no longer alive. In higher animals this function is performed chiefly by the kidneys, the lungs, and the skin, the waste matters passing off in the urine, the breath, and the sweat. In the earthworm it is principally performed by small organs called *nephridia*, of which here are two in each somite, excepting the first three or four (Fig. 29).

Each nephridium (Fig. 27) consists of a long convoluted tube, attached to the hinder face of a dissepiment, and lying in the coelom at the side of the alimentary canal. At one end the tube passes through the body-wall and opens to the exterior by a minute pore situated between the outer and inner rows of setae (p. 46). The other end of the tube passes through the dissepiment very near to the point where this is penetrated by the nerve-cord (p. 66), and opens by a broad, funnel-like expansion into the cavity of the next somite in front (*f*, Fig. 27). The margins of the funnel and the inner surface of the upper part of the tube are densely covered with powerful cilia (Fig. 28), whose action tends to produce a current setting from the coelom into the funnel and through the nephridium to the exterior.

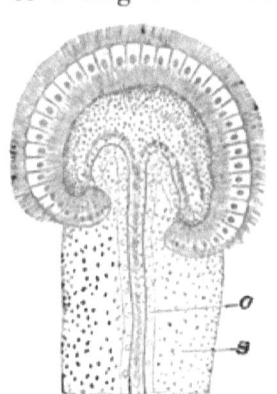

Fig. 28.—A nephridial funnel much enlarged, showing the cilia, the beginning of the ciliated canal (*c*), and the outer sheath (*s*).

The coils of the nephridium are disposed in three principal loops (I, II, III in Fig. 27). The tube itself comprises five very distinct regions, as follows :

1. The *funnel* or *nephrostome* ; much flattened from above downwards, with the opening reduced to a horizontal chink. It is composed of beautiful ciliated cells set like fan-rays around its edge. It leads into

2. The "*narrow tube*" (*n.t.*), a very delicate thin-walled contorted tube extending from the nephrostome through the first loop and a part of the second. In certain parts of its course (*a* to *b*, at *c*, and from *d* to *e*) this

tube contains cilia which are arranged in two longitudinal bands on the inner surface. At g it passes into the

3. "*Middle tube*" (*m.t.*) (g to h), extending straight through the second loop, of greater diameter, ciliated throughout, and with pigmented walls. At h it opens into the

4. "*Wide tube*" (*w.t.*). This is of still greater calibre, with granular glandular walls and without cilia. It extends through the second loop (from h to i, II) into and through the first from i to j, and finally into the third, opening at k into the

5. *Muscular part* or *duct* (*m.p.*) which forms the third loop and opens to the exterior at ex. This, the widest part of the entire nephridium, has muscular walls and forms a kind of sac or reservoir like a bladder, in which the excreted matter may accumulate and from which it may be passed out to the exterior.

The various parts of the nephridium are held together by connective tissue (p. 90), and are covered with a rich network of blood-vessels, the arrangement of which is shown in Fig. 27, B. The smaller vessels usually show numerous pouchlike dilatations which must serve to retard the flow of blood somewhat. The vessels supplying the nephridium are connected (Fig. 27, B) on the one hand with the sub-intestinal vessel through the *ventro-lateral trunks* (*v.l.*); on the other hand with the *sub-neural* (*s.n.*) and dorsal vessels, through the *dorso-tegumentary* (*d.t.*). The course of the blood is somewhat doubtful. According to the view here adopted (cf. p. 56) the blood proceeds from the dorso-tegumentary trunk to the nephridia and thence through the ventro lateral to the sub-intestinal, as shown by the arrows in the figure. Benham (from whom the figures are copied) adopts the reverse view. The development of the nephridium shows that its ciliated and glandular portions arise from a solid cord of disk-shaped cells which afterwards becomes tubular by the hollowing out of its axial portion. The tube is therefore comparable to a drain-pipe in which each cylinder represents a cell. Its cavity is not intercellular (*between* the cells, like the alimentary cavity), but *intracellular* (*within* the cells, like a vacuole).

The mode of action of the nephridia is as yet only partially understood, though there is no doubt regarding their general character. It is certain that their principal office is to remove from the body waste nitrogenous matters resulting from the decomposition of proteids; and there is reason to believe that these waste matters are passed out either as *urea* ($[NH_2]_2CO$) or as a nearly related substance, together with a certain quantity of water and inorganic salts.

Excretion in *Lumbricus* appears, however, to involve two quite distinct actions on the part of the nephridia. In the first place the glandular walls of the tube, which are richly supplied with blood-vessels, elaborate certain liquid waste substances from the blood and pass them into the cavity of the tube. In the second place the ciliated funnels are believed to take up solid waste particles floating in the cœlomic fluid and to pass them on into the tube, whence they are ultimately voided to the exterior together with the liquid products described above. It is nearly certain that these particles are derived from the breaking up of "lymphoid" cells, some of which may have been phagocytes (p. 53), floating in the cœlomic fluid, and that most if not all of these cells arise from "chloragogue cells" set free from the surface of the blood-vessels and of the intestine.

Respiration. Respiration, or breathing, is a twofold operation, consisting of the taking in of free oxygen and the giving off of carbon dioxide by gaseous diffusion through the surface of the body. Strictly speaking, this free oxygen must be regarded as food, while carbon dioxide is to be regarded as one of the excretions. Hence respiration is tributary both to alimentation and to excretion; but since many animals possess special mechanisms to carry on respiration, it is convenient and customary to treat of it as a distinct process.

Respiration is essentially an exchange of gases between the blood and the air, carried on through a delicate membrane lying between them. The earthworm represents the simplest conditions possible, since the exchange takes place all over the body, precisely as in a plant. Its moist and delicate walls are everywhere traversed by a fine network of blood-vessels lying just beneath the surface. The oxygen of the air, either in the atmosphere or dissolved in water, readily diffuses into the blood at all points, and carbon dioxide makes its exit in the reverse direction. Freed of carbon dioxide and enriched with oxygen, the blood is then carried away by the circulation to the inner parts, where it gives up its oxygen to the tissues and becomes once more laden with carbon dioxide.

In higher animals it has been proved that the red coloring matter (hæmoglobin) is the especial vehicle for the absorption and carriage of the oxygen of the blood, entering into a loose chemical union with it and readily setting it free again under the appropriate conditions. This is doubtless true in the earthworm also.

It is interesting to study the various devices by which this function is performed in different animals. In the earthworm the whole outer surface is respiratory, and no special respiratory organs exist. In other animals such organs arise simply by the differentiation of certain regions of the general surface, which then carry on the gaseous exchange for the whole organism. In many aquatic animals such regions bear filaments or flat plates or feathery processes known as *gills* or *branchiæ*, which are bathed by the water containing dissolved air, though in many such animals respiration takes place to some extent over the general surface as well. In insects the respiratory surface is confined to narrow tubes (*tracheæ*) which grow into the body from the surface and branch through every part, but must nevertheless be regarded as an infolded part of the outer surface. In man and other air-breathing vertebrates the respiratory surface is mainly confined to the lungs, which are simply localized infoldings of the outer surface specially adapted to effect a rapid exchange of gases between the blood and the air.

It is easy to see why special regions of the outer surface have in higher animals been set aside for respiration. It is essential to rapid diffusion that the respiratory surface should be covered with a thin, moist membrane, and it is no less essential that many animals should be provided with a firm outer covering as a protection against mechanical injury or desiccation. Hence the outer surface becomes more less distinctly differentiated into two parts, viz., a protecting part, the general integument; and a respiratory part, which is usually preserved from injury by being folded into the interior as in the case of lungs or tracheæ, or by being covered with folds of skin as in the gills of fishes, lobsters, etc. This covering or turning in of the respiratory surfaces brings with it the need of mechanical arrangements for pumping air or water into the respiratory chamber; and thus arise many complicated accessory respiratory mechanisms.

B. ORGANS OF RELATION. (For A see p. 49.)

Motor System. The movements of the body have a twofold purpose. In the first place they enable the animal to alter its relation to the environment, to move about (*locomotion*), to seize and swallow food, and to perform various adaptive actions in response to changes in the environment. In the second place, the movements may alter the relation of the various parts of the body one to another (*visceral movements* and the like), such as the movements which propel the blood, drive the food along the alimentary canal and roll it about (p. 49), those which expel waste matters from the nephridia, discharge the reproductive products, etc.

Most of these movements are performed by structures known as *muscles*, which consist of elongated cells (fibres) endowed in a high degree with the power of *contractility*—i.e., of shortening, or drawing together (cf. p. 27). Ordinary "muscles" are in the form of long bands or sheets of parallel fibres, such as those that form the body-wall, that move the setæ, and dilate the pharynx. Other muscular structures, however, do not form distinct "muscles," but consist of muscular fibres more or less irregularly arranged and often intermingled with other kinds of tissue. Of this character are the muscular walls of the contractile vessels, and of the muscular portions of the nephridia and dissepiments. It is clear from the above that the muscular system is not isolated, but is intimately involved in many organs.

The muscles of the body-wall are arranged in two concentric layers below the skin. In the outer layer the muscles run around the body, and are therefore called *circular* muscles. Those of the inner layers have a *longitudinal* course,—i.e., parallel with the long axis of the body,—and are arranged in a number of different bands. The most important of these are:

1. The *dorsal bands* (Fig. 39), one on either side above, in contact at the median dorsal line, and extending down on either side as far as the outer row of setæ.

2. The *ventral bands*, on either side the middle ventral line and occupying the space between the two inner (lower) rows of setæ.

3. The *lateral bands*, occupying the space on either side between the two rows of setæ.

All these vary greatly in different regions of the body, and in some parts become more or less broken up into subsidiary bands. There is also a narrow band traversing the space between the two setæ of each group.

The *setæ*, which may be reckoned as part of the motor system, are produced by glandular cells covering their inner ends, and they grow constantly from this point, somewhat as hairs grow from the root. After being fully formed, and after a certain amount of use, the setæ are cast off and replaced by new ones which have meanwhile been forming. In each group we find, therefore, setæ of different sizes. At their inner ends they are covered by a common investment of glandular cells which appears as a slight rounded prominence when viewed from within. These prominences are called the *setigerous glands*. When a worm is laid open from above, the glands are seen in four parallel rows, two of which lie on either side of the nerve-cord (see Fig. 29).

Each group of setæ is provided with special *retractor* or *protractor* muscles, and a narrow muscular band passes from the upper to the lower group on each side internal to the body-wall.

Cilia. A second set of motor organs are *cilia* (their mode of action has been referred to on p. 31), which are of the utmost importance in the life of the earthworm. They cover the inner surface of the stomach-intestine (where they doubtless assist in the movements of the food) play the important part in excretion already described, collect and help to discharge the reproductive elements (p. 74), and assist in the fertilization of the egg (p. 74). Their action, like that of the muscle-fibres, is doubtless due to the property of *contractility*, the protoplasm alternately contracting on opposite sides of the cilium and thus causing its whiplike action.

White Blood-corpuscles. Amœboid Cells. Lymph-cells. Phagocytes. Besides muscle-cells and ciliated cells there is a third variety which display contractility and movement. These are the cœlomic corpuscles referred to above (p. 53). Until recently their function was wholly unknown, but it is now generally believed that they are the scavengers of the body, devouring the dead tissues or foreign bodies which invade the organism. Whether they also attack and devour living parasites such as *Gregarina* and *Bacteria* is not yet fully determined. They move their parts much as Amœbæ do, engulfing particles about them by a kind of flux.

Nervous System. Organs of Coördination.

Introduction. The general office of the nervous system of organs is to regulate and coördinate the actions of all the other parts in such wise that these actions shall form an harmonious and orderly whole. Through nervous organs the worm receives from the environment impressions which pass inwards through the nerves as *sensory* or *afferent* impulses, to the nervous centres; and through other nervous organs impulses (*efferent* or *motor*) pass outwards from the centres to the various parts so as to arouse, modify, or suspend their activities. Thus the animal is enabled to call forth movements resulting in the two kinds of adjustments referred to on p. 62, viz., (*a*) adjustments of the body as a whole to changes in the environment (e.g., the withdrawal of the earthworm into its burrow at the approach of day); and (*b*) adjustments between the parts of the body itself, so that a change in one part may call forth answering changes in other parts (e.g., the increased supply of blood to the alimentary canal during digestion, or vigorous movements of the fore end of the body when the hind end is irritated).

These functions are always performed by one or more *nerve-cells*, which give off long slender branches known as *nerve-fibres* usually gathered together in bundles, the *nerves*, extending into all parts of the body. In all higher animals the main bulk of the nerve-cells are aggregated in definite bodies known as *ganglia*, out of which, into which, or through which, the nerves proceed; and as a matter of convenience it is customary to designate the most important of these ganglia collectively as the *cen-*

tral nervous system. The remaining portion, which consists mainly of nerve-fibres, though it may also contain many nerve-cells and small sporadic ganglia, is known as the *peripheral nervous system.*

General Anatomy of the Nervous System. In the earthworm the central system consists of a long series of double ganglia, metamerically repeated, and connected by nerve-cords known as *commissures.* The most anterior pair of ganglia, known as the *supra-œsophageal* or cerebral ganglia, lie on the dorsal aspect of the pharynx, a short distance behind the anterior extremity (Figs. 24, 29). From each of them a slender cord, the *circum-œsophageal commissure*, passes down at the side of the pharynx to end in the *sub-œsophageal* or *first ventral ganglion* on the lower side, forming with its fellow a complete ring or *pharyngeal collar* around the alimentary canal. From the sub-œsophageal ganglion a long double *ventral nerve-cord* proceeds backwards in the middle ventral line. The ventral cord consists of a series of double ganglia, one to each somite, connected by commissures and giving off lateral nerves.*

Internally the cerebral ganglia and the ventral cord (commissures as well as ganglia) consist of both nerve-cells and nerve-fibres as described on p. 94.

Peripheral Nervous System. To and from the central system just described run the nerves which constitute the peripheral system. These are as follows:

1. A pair of nerves running out on either side of each ventral ganglion and lost to view among the muscles of the body-wall.

2. A single nerve proceeding from the ventral commissures on each side immediately behind the dissepiment to which it is mainly distributed.

3. A pair of nerves from the sub-œsophageal ganglion.

4. A nerve from each half of the pharyngeal collar just beyond its divergence from its fellow. (Origin incorrectly shown.)

5. Two large cerebral nerves, which run forwards from the

---

* So closely are the two halves of the ventral cord united that its double nature can scarcely be made out without sections.

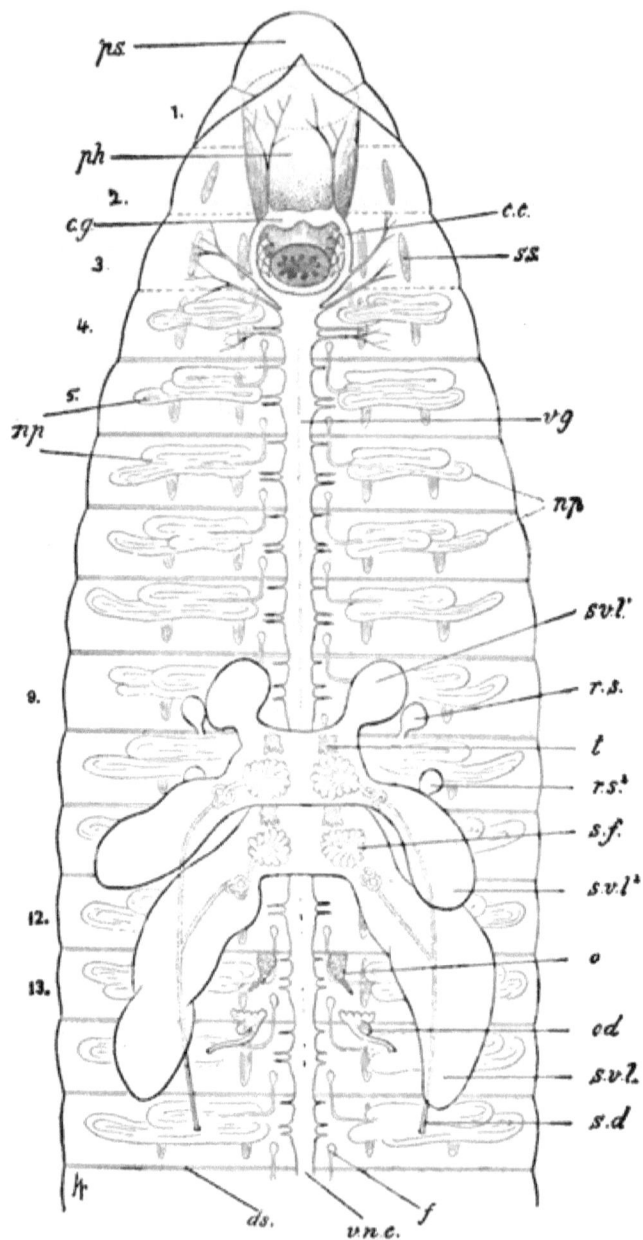

FIG. 29.—Anterior portion of the earthworm laid open from above, with the alimentary and circulatory systems dissected away. *c.c.*, circum-œsophageal commissure; *c.g.*, cerebral ganglia; *ds*, dissepiment; *f*, funnel of nephridium; *np* nephridium; *o*, ovary; *od*, oviduct; *ph*, pharynx; *ps*, prostomium; *r.s.*, seminal receptacle; *s.d.*, sperm-duct; *s.f.*, sperm-funnel; *s.v.l.*, lateral seminal vesicle; *t*, testis; *v.g.*, and *v.n.c.*, ventral nerve-cord.

cerebral ganglia, break up into many branches, and are distributed to the anterior part of the body.

Besides the main ganglia of the central system, there are many smaller ganglia in various parts of the body. Of these the most important are the *pharyngeal ganglia*—3 to 5 in number—which lie on the wall of the pharynx on each side just within the pharyngeal collar. They are connected with the latter by fine branches, and send minute nerves out upon the walls of the pharynx. This series of ganglia is often inappropriately called the *sympathetic* system.

Physiology of the Nervous System. Nerve-impulses. What is the origin and nature of a nerve-impulse? Under normal conditions the impulse is set up as the result of some disturbance, technically called a *stimulus*, acting upon the end of the fibre. A touch or pressure upon the skin, for example, acts as a stimulus to the nerve-fibres ending near the point touched—that is, it causes nerve-impulses to travel inwards along the fibres towards the central system. The nerves may be stimulated by a great variety of agents:—by mechanical disturbance, as in the case just cited, by heat, electricity, chemical action, and in special cases by waves of light or of sound, and upon this property of the nerves depends the power of the worm to receive as afferent impulses impressions from the outer world. But, besides this, nerve-fibres may also be stimulated by physiological changes taking place within the nerve-cells, which may thus send out efferent impulses to the various organs and so control their action.

Regarding the precise nature of the nerve-impulse we are ignorant, but it is probably a chemical or molecular change in the protoplasm, travelling rather rapidly along the fibre, like a wave.* We know that the nature of the impulse is not in any way dependent upon the character of the stimulus. The stimulus can only throw the nerve into action; and this action is always the same whatever be the stimulus—as the action of a clock remains the same whether it be driven by a weight or by a spring.

Co-ordination. The activities of the various organs are co-ordinated by a chain of events which in its simplest form is known as a *reflex action*, and which lies at the bottom of most of the more complicated forms of nervous action. Its nature is

---

* In the frog the nervous impulses travel at the rate of about 28 metres per second; in man it is considerably more rapid.

illustrated by the diagram (Fig. 30). Co-ordination between S and M (two organs) is not effected by a direct nervous connection, but indirectly through a nerve-centre, C, which is a nerve-cell or group of nerve-cells situated in one of the ganglia, with which both S and M are separately connected by nerve-fibres. If S be thrown into action, an *afferent* impulse travels to C, excites the nerve-centre, and causes an *efferent* impulse to travel out to M, which is thereby thrown into action also, or is modified in respect to actions already going on. Thus the actions of S and M are *co-ordinated* through the agency of C; the whole chain of events constituting a reflex action.

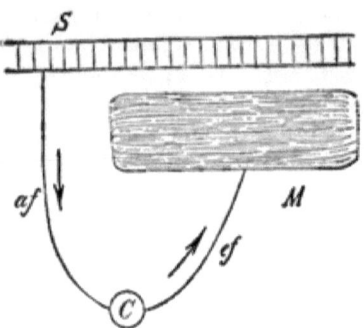

FIG. 30.—Diagram of simple reflex action. S, skin to which stimulus is applied; af, the afferent nerve-fibre; C, nerve-centre; ef, efferent nerve-fibre; M, muscle in which the efferent fibre ends.

For example, let S be the skin and M a certain group of muscles. If the skin be irritated, afferent impulses travel inwards to nerve-centres in the ganglia (C), which thereupon send forth efferent impulses to the appropriate muscles. Muscular contractions result, and the worm draws back from the unwelcome irritation.

This chain of events involves three distinct actions on the part of the nervous system which must be carefully distinguished, viz.: (*a*) the afferent impulse; (*b*) action of the centre; (*c*) the efferent impulse. It must not be supposed that the afferent impulse passes unchanged out of the centre as the efferent impulse, i.e., is simply "reflected," like a ball thrown against a wall, as the word "reflex" seems to imply. The afferent impulse as such ends with the nerve-centre, which it throws into activity. The efferent impulse is a new action set up by the agency of the centre.

There is reason to believe that many if not all nerve-centres are connected with a number of different afferent and efferent paths, and also with other centres, as shown in the diagram Fig. 31. Efferent impulses may therefore be sent out from the centre in various directions, and the precise path chosen depends on some unknown action taking place in the centre. The action of the centre moreover may be modified by efferent impulses arriving from other centres, and thus we can dimly perceive how reflexes may be controlled and guided, and how even the most complicated forms of nervous activity may be compounded out of elements similar to reflex actions.

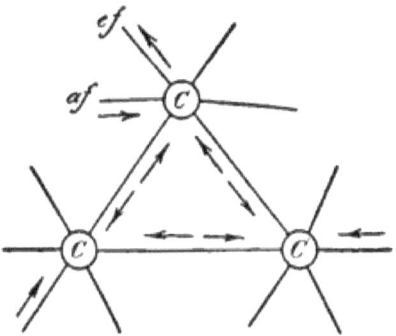

FIG. 31.—Diagram representing three nerve-centres and connections. Arrows represent the possible direction of nerve-impulses. *af*, one afferent path; *ef*, one efferent path.

There is reason to believe that in the earthworm each ventral ganglion presides over the somite to which it belongs, and is probably in the main a collection of reflex centres from whose action the element of consciousness is absent. But there is also some reason to believe that the cerebral ganglia occupy a higher position, since they probably receive the nerves of sight, taste, and smell, besides those of touch, while the ventral ganglia receive only those of touch. Experiment has shown further that the cerebral ganglia exercise to a certain limited extent a controlling action over those of the ventral chain by means of impulses sent backwards through the commissures, though this action is far less conspicuous here than in higher metameric animals such as the insects.*

The Sensitive System. (Organs of Sense.) The sensitive system is distinguished from the nervous system as a matter of convenience of description, since most of the higher animals possess definite "sense-organs" which receive stimuli and throw into action the sensory nerves proceeding from them. Although the earthworm possesses the "senses" of touch, taste, sight, and smell, it has no special organs for these senses apart from the general integument covering the surface of the body, and

---

* For a fuller discussion the student is referred to special works on Physiology.

hence can hardly be said to possess any proper sensory *system*. We do not know, moreover, whether the so-called "sensations" of the earthworm are really states of consciousness as in ourselves, for we do not even know whether earthworms possess any form of consciousness. When, therefore, we speak of the earthworm as possessing the "sense" of touch or of sight we mean simply that some of the nerves terminating in the skin may be stimulated by mechanical means or by rays of light, without necessarily implying that the worm actually feels or sees as we feel and see.

It has recently been shown that the skin contains many cells each of which gives off a single nerve-fibre that may be traced directly into the ventral nerve-cord. These "sensory cells" may be regarded as "end-organs" through which the stimuli are conveyed to the fibres. It has also been shown that these cells are aggregated in minute groups thickly scattered over the surface of the body. Each of these groups may be regarded as a simple form of sense-organ.

The sense of *touch* extends over the whole surface of the body. That of *taste* is probably located in the cavity of the mouth and pharynx; the location of the sense of *smell* is unknown. Darwin's experiments have shown that the earthworm's feeble sense of *sight* is confined to the anterior end of the body. It is probable that the nerves of sight, taste, and smell enter the cerebral ganglia alone, while those of touch run to other ganglia as well.

Systems of (Organs of) Support, Connection, Protection, etc. The structure and mode of life of many animals are such as to require some solid support to the soft parts of the body. Such supporting structures are, for instance, the bones of vertebrata, the hard outer shell of the lobster or beetle, and the coral which forms the skeleton of a polyp. The earthworm has, however, nothing of the sort, and it is obvious that a hard supporting-organ would be not only useless, but even detrimental. The power of creeping and burrowing through the earth depends upon great flexibility and extensibility of the body; and with this the presence of a skeleton might be incompatible.

The *connecting* system consists simply of various tissues by which the different organs are bound firmly together. These can only be seen upon microscopical examination. The most important of them is known as *connective tissue*.

As to *protective* structures, the earthworm is probably one of the most defenceless of animals. Nevertheless there are certain structures which are clearly for this purpose. The *cuticle* which covers the surface is a thin but tough membrane which protects the delicate skin from direct contact with hard objects. It passes into the mouth and lines the alimentary canal as far down as the beginning of the stomach-intestine. In the gizzard, where food is ground up, the cuticle is prodigiously thick and tough, and must form a very effective protection for the soft tissues beneath it. The main defence of the animal lies, however, not in any special armor, but in those instincts which lead it to lie hidden in the earth during the day and to venture forth only in the comparative safety of darkness.

CHAPTER V.

THE BIOLOGY OF AN ANIMAL (*Continued*).

The Earthworm.
Reproduction. Embryology.

Reproduction. The life of every organic species runs in regularly recurring cycles, for every individual life has its limit. In youth the constructive processes preponderate over the destructive and the organism grows. The normal adult attains a state of apparent physiological balance in which the processes of waste and repair are approximately equal. Sooner or later, however, this balance is disturbed. Even though the organism escapes every injury or special disease the constructive process falls behind the destructive, old age ensues, and the individual dies from sheer inability to live. Why the vital machine should thus wear out is a mystery, but that it has a definite cause and meaning is indicated by the familiar fact that the span of natural life varies with the species; man lives longer than the dog, the elephant longer than man.

It is a wonderful fact that living things have the power to detach from themselves portions or fragments of their own bodies endowed with fresh powers of growth and development and capable of running through the same cycle as the parent. There is therefore an unbroken material (protoplasmic) continuity from one generation to another, that forms the physical basis of inheritance, and upon which the integrity of the species depends. As far as known, living things never arise save through this process; in other words every mass of existing protoplasm is the last link in an unbroken chain that extends backward in the past to the first origin of life.

The detached portions of the parent that are to give rise to offspring are sometimes masses of cells, as in the separation of branches or buds among plants, but more commonly they are single cells, known as *germ-cells*, like the eggs of animals and the spores of ferns and mosses. Only the *germ-cells* (which may conveniently be distinguished from those forming the rest of the body, or the *somatic* cells), escape death, and that only under certain conditions.

All forms of reproduction fall under one or the other of two heads, viz., **Agamogenesis** (*asexual* reproduction) or **Gamogenesis** (*sexual* reproduction). In the former case the detached portion (which may be either a single cell or a group of cells) has the power to develop into a new individual without the influence of other living matter. In the latter, the detached portion, in this case always a single cell (ovum, oösphere, etc.), is acted upon by a second portion of living matter, likewise a single cell, which in most cases has been detached from the body of another individual. The germ is called the *female germ-cell;* the cell acting upon it the *male germ-cell;* and in the sexual process the two fuse together (*fertilization, impregnation*) to form a single new cell endowed with the power of developing into a new individual. In some organisms (e.g., the yeast-plant and bacteria) only agamogenesis has been observed; in others (e.g., vertebrates) only gamogenesis; in others still both processes take place as in many higher plants.

The earthworm is not known to multiply by any natural process of agamogenesis. It possesses in a high degree, however, the closely related power of *regeneration;* for if a worm be cut transversely into two pieces, the anterior piece will usually make good or *regenerate* the missing portion, while the posterior piece may regenerate the anterior region. Thus the worm can to a certain limited extent be artificially propagated, like a plant, by cuttings, a process closely related to true agamogenesis.* Its usual and normal mode of reproduction is by gamogenesis, that is, by the formation of male germ-cells (*spermatozoa*) and female germ-cells (*ova*). In higher animals the two kinds of germ-cells are produced by different individuals of opposite sex. The earthworm on the contrary is *hermaphrodite* or *bisexual;* every

---

* Many worms nearly related to *Lumbricus*—e.g., the genus *Dero*, and other Naids—spontaneously divide themselves into two parts each of which becomes a perfect animal. This process is true agamogenesis, though obviously closely related to regeneration.

individual is *both male and female*, producing both eggs and spermatozoa. The ova arise in special organs, the *ovaries*, the spermatozoa in *spermaries* or *testes*.

The ripe ovum (Fig. 33, *B*) is a relatively large spherical cell, agreeing closely with the egg of the star-fish (Fig. 12), but having a thinner and more delicate membrane. It is still customary to apply to ova the old terminology, calling the cell-substance *vitellus*, the membrane *vitelline* membrane, the nucleus *germinal vesicle*, and the nucleolus *germinal spot*.

The ripe spermatozoön (Fig. 33, *C*) is an extremely minute elongated cell or filament thickening towards one end to form the *head* (*n*), which contains the nucleus of the cell enveloped by a thin layer of protoplasm. This is followed by a short " middle piece " (*m*) to which is attached a long vibratory flagellum or tail (*t*). The tail is virtually a long cilium (p. 31), which by vigorous lashing drives the whole cell along head-foremost, very much as a tadpole is driven by its tail.

Since the ovaries and spermaries give rise to the germ-cells, they are called the *essential organs* of reproduction. Besides these, *Lumbricus*, like most animals, has *accessory organs* of reproduction which act as reservoirs or carriers of the germs, assist in securing cross-fertilization, and minister to the wants of the young worms.

Essential Reproductive Organs. The *ovaries* are two in number and lie one on either side in the 13th somite attached to the hinder face of the anterior dissepiment (*ov*, Fig. 29). They are about 2^{mm} in length, distinctly pear-shaped, and attached by the broader end (Fig. 32). The narrow extremity contains a single row of ova and is called the *egg-string* (*es*). In this the ova are ripe or nearly so; behind they shade off into those more and more immature, till these are lost in a mass of nearly undifferentiated cells (*primitive ova*), constituting the great bulk of the ovary. Each of these,

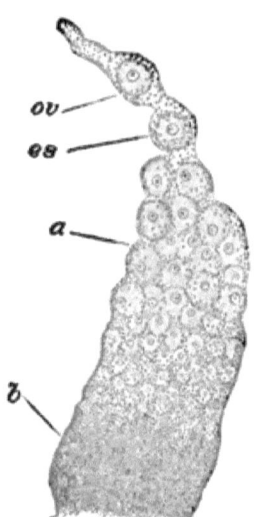

FIG. 32.—The ovary, much enlarged. *b*, the basal part; *a*, body of the ovary containing immature ova; *es*, egg-string; *ov*, ripe ovum ready to fall off.

however, is surrounded with still smaller cells constituting its nutrient envelope or *follicle*. As the ova mature the follicles still persist, and they may be detected even in the eggstring. When fully ripe the ovum bursts the follicle and is shed from the end of the egg-string into the body-cavity. It is ultimately taken into the oviduct and carried to the exterior.

The development of the ovary shows it to be morphologically a thickening of the peritoneal epithelium. The eggs therefore are originally epithelial cells.

The *spermaries* or *testes* (t,t, Fig. 29) are four in number and in outward appearance are somewhat similar to the ovaries. They are small flattened bodies with somewhat irregular or lobed borders, lying one on either side the nerve-cord in a position corresponding with that of the ovaries, but in somites 10 and 11. Like the ovary the testis is a solid mass of cells, which are shed into the body-cavity and are finally carried to the exterior. The sperm-cells leave the testis, however, at a very early period and undergo the later stages of maturation within the cavities of the seminal vesicles described below.

Accessory Reproductive Organs. The most important of the accessory organs are the genital *ducts*, by which the germ-cells are passed out to the exterior. Both the female ducts (*oviducts*) and the male (*sperm-ducts*) are tubular organs opening at one end to the outside, through the body-wall, and at the other end into the cœlom by means of a ciliated funnel somewhat similar to a nephridial funnel, but much larger. By means of these ciliated funnels the germ-cells after their discharge from the ovary or testis are taken up and passed to the exterior.

The oviducts (*od*, Fig. 29, Fig. 23) are two short trumpet-shaped tubes lying immediately posterior to the ovaries and passing through the dissepiment between the 13th and 14th somites. The inner end opens freely into the cavity of the 13th somite, by means of a wide and much-folded ciliated funnel, from the centre of which a slender tube passes backward through the dissepiment, turns rather sharply towards the outer side and, passing through the body-wall, opens to the outside on the 14th somite (see p. 43). Immediately behind the dissepiment the oviduct gives off at its dorsal and outer side a small pouch, richly supplied with blood-vessels. In this, the *receptaculum ororum*, the ova taken up by the funnel are temporarily stored before passing out to the exterior.

It is probable that the eggs never float freely in the cœlom, but drop out of the ovary at maturity directly into the mouth of the funnel. They pass thence into the *receptaculum*, where they may remain for a considerable period.

The *sperm-ducts* (*vasa deferentia*) (*sd*, Fig. 29) are very long slender tubes, open like the oviducts at both ends. The outer opening is a conspicuous slit surrounded by fleshy lips (Fig. 21), on the ventral side of the 15th somite. From this point the duct runs straight forwards to the 12th somite, where it branches like a Y, the two branches passing forwards to terminate, one in the 11th somite, the other in the 10th. Near its end each branch is twisted into a peculiar knot and finally terminates in an immense ciliated funnel (the so-called "ciliated rosette"), the borders of which are folded in so complicated a manner that they form a labyrinthine body, the true nature of which can only be made out in microscopic sections.

The two pairs of sperm-funnels (Fig. 29) lie in the 10th and 11th somites, immediately posterior to the respective testes, i.e., they have essentially the same relation to the testes as that of the oviduct-funnels to the ovaries.

The testes and sperm-funnels can be readily made out only in young specimens. In mature worms they are completely enveloped by the seminal vesicles described below.

Seminal vesicles. These, the most conspicuous part of the reproductive apparatus, are voluminous pouches in which the sperm-cells undergo their later development, after leaving the testis. They are large white bodies lying in somites 9 to 12 and usually overlapping the œsophagus in that region. In all cases there are three pairs of lateral seminal vesicles, viz., an anterior pair in somite 9, a middle pair in somite 11, and a posterior pair in somite 12. In immature specimens these six are entirely separate, and allow the testes to be easily seen. In mature worms (as shown in Fig. 29) the posterior pair of lateral vesicles grow together in the middle line, thus forming a *posterior median vesicle* lying below the alimentary canal in the 11th somite. In like manner an *anterior median vesicle* is formed in the 10th somite by the union of the *two* anterior pairs of lateral vesicles. The two median vesicles thus formed envelop the testes and sperm-funnels of their respective somites and hide them from view.

The sperm-cells leave the testis at a very early period and float freely in the cavities of the seminal vesicles, where many stages of their development may easily be observed. They are developed in balls known as *spermatospheres*, each of which consists of a central solid mass of protoplasm surrounded by a single layer of sperm-cells. When mature the spermatozoa separate from the central mass and are drawn into the funnels of the sperm-ducts. The manner in which this action is controlled is not understood.

The *seminal receptacles* are accessory organs of reproduction in the shape of small rounded sacs or pouches, open to the outside only, at about the level of the upper row of setæ. They lie between the 9th and 10th, and 10th and 11th somites (*s.r*, Figs. 24 and 29), where their openings may be sought for (Fig. 21). Their function is explained under the head of copulation.

Accessory glands. Besides all the structures so far described there are many glands which play a part in the reproductive functions. The setigerous glands from about the 7th to about the 19th somite (sometimes fewer, sometimes none at all) are often greatly enlarged, and form the glandular prominences mentioned at p. 46. They seem to be used as organs of adhesion during copulation. The clitellum is filled with gland-cells which probably serve in part to secrete a nourishing fluid for the young worms, and in part to provide a tough protecting membrane to cover them.

Copulation. Egg-laying. Inasmuch as each individual earthworm produces both ova and spermatozoa, it might be supposed that copulation, or the sexual union of two different individuals, would not be necessary. This, however, is not the case. The ova of one individual are invariably fertilized by the spermatozoa of another individual after a process of copulation and exchange of spermatozoa, as follows: During the night-time, and usually in the spring, the worms leave their burrows and pair, placing themselves so that their heads point in opposite directions and holding firmly together by the enlarged setigerous glands and the thickened lower lateral margins of the clitellum. During this act the seminal receptacles of each worm are filled with spermatozoa from the sperm-ducts of the other, after which the worms separate. [The spermatozoa thus received are simply stored up and do not perform their function until the time of egg-laying.]

When the worm is ready to lay its eggs the glands of the clitellum become very active, pouring out a thick glairy fluid which soon hardens into a tough membrane and forms a girdle around the body. Besides this a large quantity of a thick jelly-like nutrient fluid is poured out and retained in the space between the girdle and the body of the worm. The girdle is thereupon gradually worked forward toward the head of the worm by contractions of the body. As it passes the 14th somite a number of ova are received from the oviducts, and between the 9th and 11th somites a quantity of spermatozoa are added from the seminal receptacles where they have been stored since the time of copulation, when they were obtained from another worm. The girdle is next stripped forwards over the anterior end and is finally thrown completely off. As it passes off its open ends immediately contract tightly together, and the girdle becomes a closed capsule (Fig. 33) containing both ova and spermatozoa floating in a nutritive fluid or milk. The membrane soon assumes a light yellowish or brown color, becomes hard and tough, and serves to protect the developing embryos. The capsules may be found in May or June in earth under logs or stones, or especially in heaps of manure. Within the capsules the fertilization and development of the ova take place.

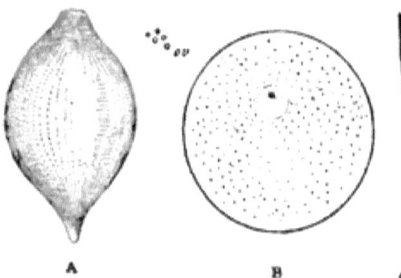

FIG. 33.—A, egg-capsule enlarged 5 diameters (a few eggs, or, enlarged to the same scale are shown near by on the right); B, an ovum very much enlarged; C, a spermatozoön, enormously magnified; n, head; m, middle piece; t, tail.

Fertilization and Embryological Development. The spermatozoa swim actively about in the nutrient fluid of the capsule, approach an ovum, and attach themselves to its surface by their heads. Several of the spermatozoa then enter the vitellus (cf. p. 80), but it has been proved that only one of these is concerned in fertilization, the others dying and becoming absorbed by the ovum.

FERTILIZATION OF THE EGG. 79

It is probable that the tail plays no part in the actual fertilization, but is merely a locomotor apparatus for the head (nucleus) and middle-piece.

Within the ovum the head of the spermatozoön persists as the *sperm-nucleus* (or *male pro-nucleus*), while the protoplasm in its neighborhood assumes a peculiar and characteristic radiate arrangement like a star, probably through the influence of the middle-piece.

After the entrance of the spermatozoön the egg segments off

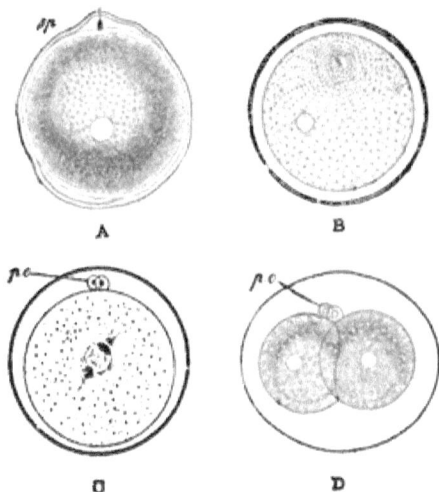

FIG. 34.—Fertilization of the ovum. *A*, entrance of the spermatozoön (in the sea-urchin, after Fol). *B*, the sea-urchin egg after entrance of the spermatozoön; within and to the left is the egg-nucleus; above is the sperm-nucleus, with a centrosome near it (modified from Hertwig). *C*, diagram of the ovum after extrusion of the polar cells (*p.c.*), and union of the two pro-nuclei to form the segmentation-nucleus. The smaller and darker portion of the latter is derived from the sperm-nucleus. Two asters or archoplasm-spheres are shown near the nucleus. These arise by the division of a single aster derived from the middle-piece of the spermatozoön. *D*, two-celled stage of the earthworm, after the first fission of the ovum. (After Vejdovsky.)

at one side two small cells, one after the other, known as the *polar cells* or *polar bodies*. These take no part in the formation of the embryo, and their formation probably serves, in some way not yet wholly clear, to prepare the egg for the last act of fertilization. After the formation of the polar cells the egg-nucleus (now often called the *female pro-nucleus*) and the sperm-nucleus approach one another and finally become intimately associated to form the *segmentation-* or *cleavage-nucleus ;* by this act fertilization is completed.

The process of fertilization appears to be essentially the same among all higher animals, and in a broader sense to be identical with the sexual process among all higher and many lower plants (compare the fern, p. 139), but its precise nature is still in dispute. It is certain that one essential part of it is the union of two nuclei derived from the two respective parents. This has led to the view, now held by many investigators, that inheritance has its seat in the nucleus, and that chromatin (p. 23), is its physical basis. Later researches have shown that another element known as the archoplasm- or attraction-sphere is concerned in fertilization, and this is apparently always derived from the middle-piece. It is not yet certain whether the archoplasm is to be regarded as a nuclear or a cytoplasmic structure, and it is equally doubtful whether it plays an essential or merely a subsidiary rôle in fertilization and inheritance (cf. p. 84).

Cleavage of the Fertilized Ovum. Soon after fertilization the ovum begins the remarkable process of segmentation which has already been briefly sketched on p. 25. The segmentation-nucleus divides into two parts, and this is followed by a division of the vitellus, each half of the original nucleus becoming the nucleus of one of the halves of the vitellus; that is, the original cell divides into two smaller but similar cells (see Fig. 34). These divide in turn into four, and these into eight, and so on, but yet remain closely connected in one mass. In the case of the earthworm, the cells do not multiply in regular geometrical progression, but show many irregularities; and moreover they become unequal in size at an early period.

The blastula (pp. 25, 85,) shows scarcely any differentiation of parts, though the cells of one hemisphere are somewhat smaller than the others. From this time forwards the whole course of development is a process of differentiation, both of the cells and of the organs into which they soon arrange themselves. One of the first steps in this process is a flattening of the embryo at the lower pole—i.e., the half consisting of larger cells (Fig. 35, *D*). The large cells are then folded into the segmentation-cavity so as to form a pouch opening to the exterior; at the same time the embryo becomes somewhat elongated (Fig. 35, *E, F*).

This process is known as *gastrulation*, and at its completion the embryo is called the *gastrula*. The infolded pouch (called the *archenteron*) is the future alimentary canal; its opening (now known as the *blastopore*) will become the mouth; and the layer of small cells over the outside will form the skin or outer layer of the body-wall.

The embryo very soon begins to swallow, through the blastopore, the milklike fluid in which it floats, and to digest it within the cavity of the archenteron.

It is obvious that the embryo already shows a distinct differ-

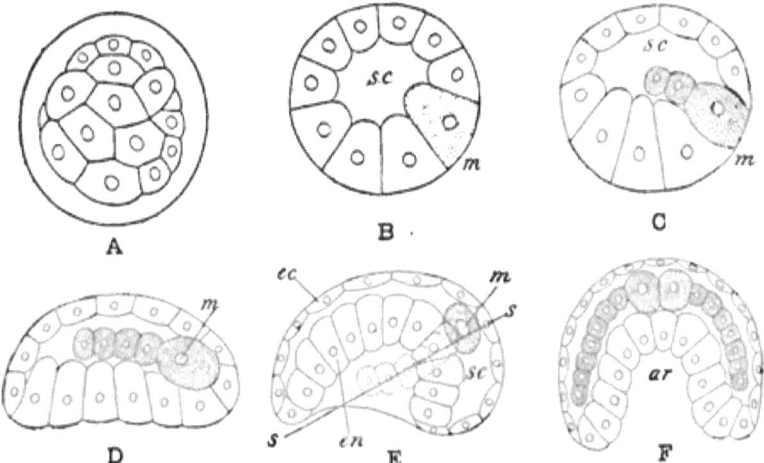

FIG. 35.—Diagrams of the early stages of development in the earthworm. *A*, accurate drawing of the blastula, surrounded by the vitelline membrane (after Vejdovsky); *B*, blastula in optical section showing the large segmentation-cavity (*s,c*), and the parent-cell of the mesoblast (*m*); *C*, later blastula, showing formation of mesoblast-cells; *D*, flattening of the blastula preparatory to invagination; *E*, the gastrula in side view; as the infolding takes place the two mesoblast-bands are left at the sides of the body, in the position shown by the dotted lines; *F*, section of *E* along the line *s-s*, showing the mesoblast-bands and pole-cells.

entiation of parts which perform unlike functions. In fact we may regard the gastrula as composed of two tissues still nearly similar in structure though unlike in function. One of these consists of the layer of cells which forms the outer covering; this tissue is known as the *ectoblast* (*ec*, Fig. 35). The second tissue is the layer of cells forming the wall of the archenteron; it is called the *entoblast* (*en*). The ectoblast and entoblast together are known as the *primary germ-layers*.

Meanwhile changes are taking place which result in the formation of a third germ-layer lying in the segmentation-cavity between the ectoblast and entoblast and therefore called the *mesoblast* (*m*, Figs. 35, 36). In some animals the mesoblast does not arise until after the completion of gastrulation. In

Lumbricus, however, it goes on during gastrulation and begins even before gastrulation. Even in the blastula stage two large cells may be distinguished which afterwards give rise to the mesoblast and are hence called the *primary mesoblastic cells*. They soon bud forth smaller cells into the segmentation-cavity, and as the blastula flattens they themselves sink below the surface. At this period, therefore, the mesoblast forms two bands of cells (*mesoblast-bands*) each terminating behind in the large mother-cell or *pole-cell*. Throughout the later stages the pole-cells continue to bud forth smaller cells which are added to the hinder ends of the mesoblast-bands (Figs. 35, 36).

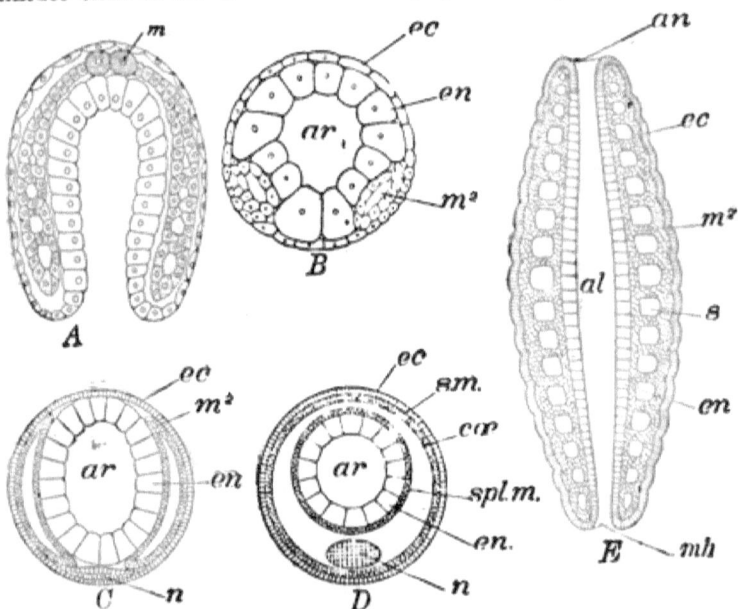

FIG. 36.—Diagrams of later embryonic stages. *A*, late stage in longitudinal section, showing the appearance of the cavities of the somites; *B*, the same in cross-section; *E*, diagram of a young worm in longitudinal section after the formation of the stomodæum, proctodæum, and anus; *C*, the same in cross-section, showing the beginning of the nervous system; *D*, cross-section of later stage with the nervous system completely established. *al*, alimentary canal; *ar*, archenteron; *an*, anus; *cœ*, cœlom; *ec*, ectoblast; *en*, entoblast; *m¹*, primary mesoblastic cells; *m²*, mesoblast; *mh*, mouth; *n*, nervous system; *s*, cavity of somite; *s.m*, somatic layer of the mesoblast, which with the ectoblast forms the somatopleure; *spl.m*, splanchnic layer of the mesoblast, which with the entoblast forms the splanchnopleure.

After each division the pole-cells increase in size, so that up to a late stage in development they may be distinguished from

CELL-DIVISION. KARYOKINESIS.

the cells to which they give rise. The two masses of mesoblastic cells gradually increase in size and finally fill the segmentation-cavity.

The internal phenomena of cell-division are of great complexity and can here be given only in outline. The ordinary type of cell-division, as shown in the segmentation of the ovum and in the multiplication of most tissue-cells, involves a complicated series of changes in the nucleus known as *karyokinesis* or *mitosis*. These changes, which appear to be of essentially the same character in nearly all kinds of cells, and both in plants and in animals, are illustrated by the following diagrams:

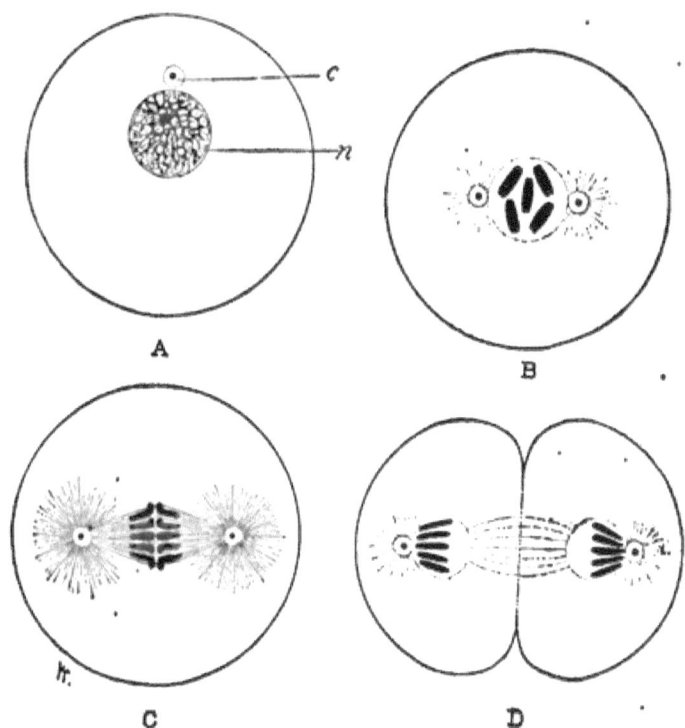

FIG. 37.—Diagrams of indirect cell-division or karyokinesis.
  A. Cell just prior to division, showing nucleus (*n*) with its chromatic reticulum and the attraction-sphere and centrosome (*c*).
  B. First phase; the attraction-sphere has divided into two, which have moved 180° apart; the reticulum has been resolved into five chromosomes (black), each of which has split lengthwise.
  C. Second phase; fully developed karyokinetic figure (*amphiaster*), with spindle and asters; the chromosome-halves are moving apart.
  D. Final phase; the cell-body is dividing, the spindle disappearing, the daughter-nuclei about to be formed.

In its resting state the nucleus contains a network or *reticulum* of chromatin (Fig. 37, *A*). As the cell prepares for division a small body (*c*) makes its appearance near the nucleus, known as the *attraction-sphere* or *archoplasm-mass*, and in its interior there is often a smaller body, the *centrosome*. The first step in cell-division is the fission of the archoplasm-mass into two, each containing a centrosome (derived by fission of the original centrosome); after this the two masses move apart to opposite poles of the nucleus (Fig. 37, *B*). The reticulum now becomes, in most cases, resolved into a thread coiled into a skein (not shown in the figure), which finally breaks up into a number of bodies known as *chromosomes*. Their form (granular, rodlike, loop-shaped) and number (two, eight, twelve, sixteen, etc., or often much higher numbers) appear to be constant for each species of plant and animal. The second principal step is the longitudinal splitting of each chromosome into halves (Fig. 37, *B*) and the disappearance of the nuclear membrane.

In the third place starlike rays (*aster*) appear in the protoplasm around the archoplasm-masses, a spindle-shaped structure appears between them (Fig. 37, *C*), and the double chromosomes arrange themselves around the equator of the spindle. The structure thus formed is known as the *amphiaster* or *karyokinetic figure*.

Fourthly, the two halves of each chromosome move apart towards the respective poles of the spindle and the entire cell-body then divides in a plane passing through the equator of the spindle. Each group of daughter-chromosomes now gives rise to a reticulum, which becomes surrounded with a membrane and forms the nucleus of the daughter-cell. The spindle disappears, and in some cases the archoplasm-mass, with its star-rays (aster), seems to disappear also. In other cases, however, the archoplasm-mass and centrosome persist and may be found in the resting cell (e.g., in leucocytes and connective-tissue cells), lying near the nucleus in the cytoplasm.

It appears from the foregoing description that each daughter-cell receives exactly half the substance of the mother-nucleus (chromatin), mother-archoplasm, and mother-centrosome. In many cases the cytoplasm also divides equally, in other cases unequally.

It has been proved in a considerable number of cases that in the fertilization of the ovum each germ-cell contributes the same number of chromosomes, and the wonderful fact has been established with high probability that the paternal and maternal chromatic substances are equally distributed to the two cells found at the first segmentation of the ovum. It is further probable that this equal distribution continues in all the later divisions; and if this is true, *every cell in the whole adult body contains material directly derived from both parents, and hence may inherit from both.*

Gastrulation. Germ-layers. Differentiation. Origin of the Body. Almost from the first the cells arrange themselves so as to surround a central cavity known as the *segmentation-cavity*. This cavity increases in size in later stages, so that the embryo finally appears as a hollow sphere surrounded by a wall consisting of a single layer of cells. This stage is known as the *blastula* (or *blastosphere*) (*A*, *B*, Fig. 35).

The formation of the GERM-LAYERS is one of the most important and significant processes in the whole course of development. Germ-layers like those of *Lumbricus*, and called by the same names, are found in the embryos of all higher animals; and it will hereafter appear that this fact has a profound meaning.

Development of the Organs. (**Organogeny.**) The embryo gradually increases in size and at the same time elongates. As it lengthens, the blastopore (in this case the *mouth*) remains at one end, which is therefore to be regarded as anterior, and the elongation is backwards. The cells of all three germ-layers continually increase in number by division, new matter and energy being supplied from the food, which is swallowed by the embryo in such quantities as to swell up the body like a bladder. The archenteron enlarges until it comes into contact with the ectoblast and the segmentation-cavity is obliterated.

The two primary mesoblastic cells are carried backwards, and always remain at the extreme posterior end (*m*, Fig. 36). The mesoblast is in the form of two bands lying on either side of the archenteron, and extending forwards from the primary mesoblastic cells.

This is clearly seen in a cross-section of the embryo, as in Fig. 36, *B*, *C*. The mesoblastic bands are at first solid, but after a time a series of paired cavities appears in them, continually increasing in number by the formation of new cavities near the hinder end of the bands as they increase in length. A cross-section passing through one pair of these cavities is shown at *B*, Fig. 35. As the bands lengthen they also extend upwards and downwards (*C*, Fig. 35), until finally they meet above and below the archenteron. The cavities at the same time continue to increase in size, and finally meet above and below the archenteron, which thus becomes surrounded by the body-cavity or cœlom (*D*). The cavities are separated by the double partition-walls of mesoblast. These partitions are the dissepiments, and the cavities themselves constitute the cœlom. The outer mesoblastic wall of each cavity is known as the *somatic layer* (*s.m.*); it unites with the ectoblast to constitute the body-wall (*somatopleure*). The inner wall, or *splanchnic layer* (*spl.m*), unites with the entoblast to constitute the wall of the alimentary canal (*splanchnopleure*). An ingrowth of ectoblast (*stomodæum*) takes place into the blastopore to form the pharynx, and a similar ingrowth at the opposite extremity (*proctodæum*) unites with the blind end of the archenteron to form the anus and terminal part of the intestine.

As to its origin, therefore, the alimentary canal consists of three portions, viz. : (1) the archenteron, consisting of the

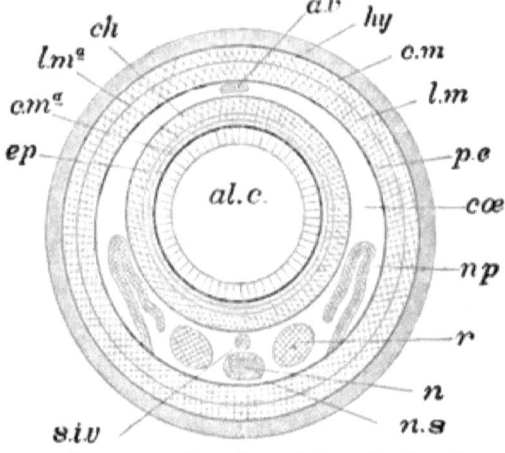

FIG. 38.—Diagram of a cross-section of *Lumbricus*, showing the relation of the various organs, etc., to the germ-layers. Ectoblastic structures shaded with fine parallel lines, entoblastic with coarser parallel lines, mesoblastic with cross-lines; *al.c*, alimentary canals; *ch*, chloragogue layer; *cœ*, cœlom; *c.m*, circular muscles of body-wall; *c.ma*, circular muscles of alimentary wall; *ep*, lining epithelium of alimentary canal; *d.v*, dorsal vessel; *hy*, hypodermis or skin; *l.m*, longitudinal muscles of body-wall; *l.m.a*, longitudinal muscles of alimentary wall; *n*, central part of nerve-cord; *np*, nephridium; *n.s*, sheath of nerve-cord; *p.e*, peritoneal epithelium; *r*, reproductive organs; *s.i.v*, sub-intestinal vessel.

original entoblast; (2) the stomodæum or pharyngeal region, lined by ectoblast; and (3) the proctodæum or hindmost part, also lined by ectoblast. These three parts are called the *fore-gut* (stomodæum), *mid-gut* or mensenteron (archenteron), and *hind-gut* (proctodæum), and it is a remarkable fact that these same parts can be distinguished in all higher animals, not excepting man.

The body now becomes jointed by the appearance of transverse folds opposite the dissepiments, and the metamerism of the body becomes evident on the exterior. The young worm has thus reached a stage (*E*, Fig. 36) where its resemblance to the adult is obvious. It has an elongated, jointed body, traversed by the alimentary canal, which opens in front by the mouth and behind by the anus. The metamerism is expressed externally by the jointed appearance, internally by the presence of paired cavities (cœlom) separated by dissepiments. Both the body-wall and the alimentary wall consist of two layers: the former of ectoblast without and somatic mesoblast within; the latter of splanchnic mesoblast without (i.e., towards the body-cavity), and either entoblast or ectoblast within, according as we consider the mid-gut on the one hand, or the fore- and hind-gut on the other. This is shown in Fig. 38, which represents a cross-section of the embryo through the mid-gut. If this be clearly borne in mind the development of all the other organs is easy to understand, since they are formed as thickenings, outgrowths, etc., of the parts already existing. For instance, the blood-vessels make their appearance everywhere throughout the mesoblast, and the reproductive organs are at first mere thickenings on the somatic layer of the mesoblast, afterwards separating more or less from it so as to lie in the cavity of the cœlom. The nervous system is produced by thickenings and ingrowths from the ectoblast. The origin of the different parts is shown in the following scheme:—

THE GERM-LAYERS AND THEIR DERIVATIVES.

Ectoblast.	Outer skin (Hypodermis and Cuticle). Nerves and Ganglia. Lining membrane of pharynx (fore-gut). Lining membrane of anus and hinder part of intestine (hind-gut).
Mesoblast.	Muscles. Blood-vessels. Reproductive organs. Outer layers of alimentary canal.
Entoblast.	Lining membrane of greater part of the alimentary canal (mid-gut).

The above statements * as to the origin of the various organs acquire great interest in view of the fact that they are essen-

---

* The nephridia have been omitted since their precise origin is in dispute. It is certain that the outer portion of the tube (muscular part) is an ingrowth from the ectoblast. The latest researches seem to show that the entire nephridium has the same origin, though some authors describe the inner portion as arising from mesoblast.

tially true of all animals above the earthworm, as well as of many below it—of all, in a word, in which the three germ-layers are developed, i.e., all those above the *Cœlenterata*, or polyps, jelly-fishes, hydroids, sponges, etc. In man, as in the earthworm and all intermediate forms, the ectoblast gives rise to the outer skin (epidermis), the brain and nerves, fore- and hind-gut; the entoblast gives rise to the lining membrane of the stomach, intestines, and other parts pertaining to the mid-gut; while the somatic and splanchnic layers of the mesoblast give rise to the muscles, kidneys, reproductive organs, heart, blood-vessels, etc. It is now generally held that the germ-layers throughout the animal kingdom (with the partial exception of the *Cœlenterata* already mentioned) are essentially identical in origin and fate. This view is known as the *Germ-layer Theory*. It is one of the most significant and important generalizations which the study of Embryology has brought to light, since it recognizes a structural identity of the most fundamental kind among all the higher animals.

Sooner or later the young earthworm bursts through the walls of the capsule and makes its entry into the world. When first hatched it is about an inch long and has no clitellum.

It is a curious fact that in certain species of *Lumbricus* the young worms are almost always hatched as *twins*, two individuals being derived from a single egg by a process which is described by Kleinenberg in the *Quarterly Journal of Microscopical Science*, Vol. XIX., 1879. It often happens that the twins are permanently united by a band of tissue, as in the case of the well-known Siamese twins.

We have now traced roughly the evolution of a complex many-celled animal from a simple one-celled germ. It is important to notice at this point a few general principles which are true of higher animals in general.

1. The embryological history is a true process of development,—not a mere growth or unfolding of a pre-existing rudiment as the leaf is unfolded from the bud. Neither the ovum nor any of the earlier stages of development bears the slightest resemblance to an earthworm. The embryo undergoes a transformation of structure as well as an increase of size.

2. It is a progress from a one-celled to a many-celled condition.

3. It is a progress from relative simplicity to relative complexity. The ovum is certainly vastly more complex than it appears to the eye, but no one can doubt that the full-grown worm is more complex still.

4. It is a progress from a slightly differentiated to a highly differentiated condition. The life of the ovum is that of a single cell. The blastula is composed of a number of nearly similar cells, which in the gastrula become differentiated into two distinct tissues. In later stages the cells become differentiated into many different tissues, which in turn build up different organs performing unlike functions.

5. Lastly, the development forms a cycle, beginning with the germ-cell, and after many complicated changes resulting in the production of new germ-cells, which repeat the process and give rise to a new generation. All other cells in the body must sooner or later die. The germ-cells alone persist as the starting-point to which the cycle of life continually returns (cf. p. 73). Their protoplasm, the "*germ-plasm*," is the bond of continuity that links together the successive generations.

CHAPTER VI.

THE BIOLOGY OF AN ANIMAL (*Continued*).

The Earthworm.

MICROSCOPIC STRUCTURE OR HISTOLOGY.

WE have followed the development of the one-celled germ through a stage, the *blastula*, in which it consists of a mass of nearly similar cells out of which the various tissues of the adult eventually arise. The first step in this direction is the differentiation of the *germ-layers* or three primitive tissues (p. 84). As the embryo develops, the cells of these three tissues become *differentiated in structure* to fit them for different duties in the physiological division of labor. And when this process of differentiation is accomplished and the adult state is reached we find six well-marked varieties of tissue, as follows:—

PRINCIPAL TISSUES OF *Lumbricus*.

I. **Epithelial.** Layer of cells covering free surfaces.
    (*a*) *Pavement Epithelium.* Cells thin and flat, arranged like the stones of a pavement.
    (*b*) *Columnar Epithelium.* Cells elongated, standing side by side, palisade-like.
    (*c*) *Ciliated Epithelium.* Columnar or cuboid, and bearing cilia.

II. **Muscular.** Cells contractile and elongated to form *fibres*. Often arranged in parallel masses or *bundles*.

III. **Nervous.** Cells pear-shaped or irregular, with large nuclei; having processes prolonged into slender cords or fibres, bundles of which constitute the *nerves*.

IV. **Germinal.** Including the germ-cells. At first in the form of epithelial cells covering the cœlomic surface, but afterwards differentiated into ova and spermatozoa.

V. **Blood.** Isolated cells or corpuscles floating in a fluid intercellular substance, the *plasma*.

VI. **Connective Tissue.** Cells of different shapes, often branched but sometimes rounded, separated from one another by more or less lifeless (intercellular) substance in the form of threads or homogeneous material.

These six kinds of tissue constitute the main bulk of the earthworm, as of higher animals generally; but there are in addition other tissues which will be treated of hereafter.

Arrangement of the Tissues. The simplest and most direct mode of discovering the arrangement of the tissues is by the microscopical study of thin transverse or longitudinal sections. A

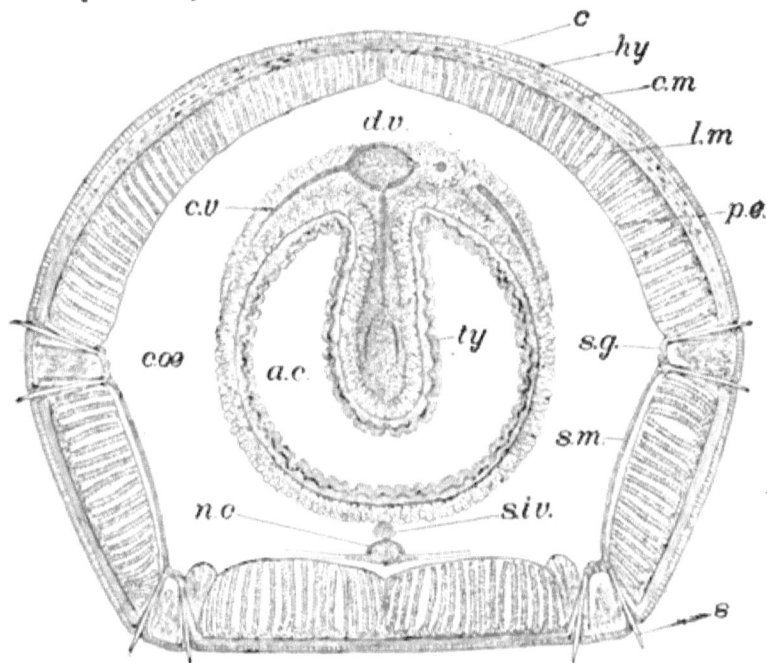

FIG. 39.—Transverse section of the body behind the clitellum. *a.c*, cavity of the alimentary canal; *c*, cuticle; *cœ*, cœlom; *c.m*, circular muscles; *c.v*, circular vessel; *d.v*, dorsal vessel; *hy*, hypodermis; *l.m*, longitudinal muscles; *n.c*, ventral nerve-chain; *p.e*, peritoneal epithelium; *s*, seta; *s.g*, setigerous gland; *s.i.v*, sub-intestinal vessel; *s.m*, muscle connecting the two groups of setæ on the same side; *ty*, typhlosole.

transverse section taken through the region of the stomach-intestine is represented in Fig. 39. Its composition is as follows:—

A. BODY-WALL.

This consists of five layers, viz. (beginning with the outside),—

1. *Cuticle* (*c*). A very thin transparent membrane, not composed of cells and perforated by fine pores. It is a product or secretion of the—

2. *Hypodermis* (*hy*) (epidermis **or skin**). A layer of columnar epithelium, composed of several kinds of elongated cells, set vertically to the surface of the body. Some of these, known as *gland-cells*, have the power of producing within their substance a glairy fluid (mucus), which exudes to the exterior through the pores in the cuticle. Others (sensory cells) give off from their inner ends nerve-fibres which may be traced inwards to the ganglia (Fig. 43).

The Clitellum is produced by an enormous thickening of the hypodermis, caused especially by a great development of the gland-cells. Three forms of these may be distinguished, which probably produce different secretions. The tissue is permeated by numerous minute blood-vessels which ramify between the cells.

3. *Circular Muscles* (*c.m*). A layer of parallel muscle-fibres running around the body. On the upper side they are intermingled with connective-tissue cells containing a granular brownish substance (pigment) which gives to the dorsal aspect its darker tint.

4. *Longitudinal Muscles* (*l.m*). A layer of muscle-fibres running lengthwise of the body. They are arranged in complicated bundles, which in cross-sections have a feathery appearance. In longitudinal sections they appear as a simple layer, and resemble the circular fibres as seen in the cross-section.

The circular muscles are arranged in somewhat similar bundles, as may be seen in longitudinal sections.

5. *Cœlomic or Peritoneal Epithelium* (*p.e*). A very thin layer of flattened cells next the cœlomic cavity.

The hypodermis, and therefore also the cuticle to which it gives rise, is derived from the ectoblast. The other layers (3, 4, 5) arise from the somatic layer of the mesoblast.

B. ALIMENTARY CANAL.

The wall of this tube appears in cross-section as a ring surrounded by the cœlom. The typhlosole (*ty*) is seen to be a deep infolding of its upper portion. In the middle region the wall is composed of five layers as follows, starting from the alimentary cavity (Fig. 40):—

1. *Lining Epithelium* (*ep*). A layer of closely packed, narrow ciliated columnar cells with oval nuclei.

2. *Vascular Layer* (*v.l*). Numerous minute blood-vessels.

3. *Circular Muscles* (*c.m*). A thin layer of muscle-fibres running around the gut.

4. *Longitudinal Muscles* (*l.m*). A thin layer of muscle-fibres running along the gut.

5. *Chloragogue Layer* (*ch*). Composed of large polyhedral or rounded cells containing yellowish-green granules. The cells fill the hollow of the typhlosole, and cover the surface of the dorsal and lateral blood-vessels. This layer represents the splanchnic part of the peritoneal epithelium.

The same general arrangement exists in all parts of the alimentary canal, but is sometimes greatly modified. For instance, the gizzard and pharynx are lined by a tough, thick cuticle, and the muscular layers are enormously developed. In a part of the gizzard the chloragogue-layer is nearly or quite absent and the typhlosole disappears. A fuller description of these modifications will be found in Brooks's *Handbook of Invertebrate Zoology*, and a complete account in Claparède, *Zeitschrift für wissenschaftliche Zoologie*, Vol. XIX., 1869.

The lining epithelium is derived from the entoblast. The remaining layers arise by differentiation of the splanchnic layer of mesoblast.

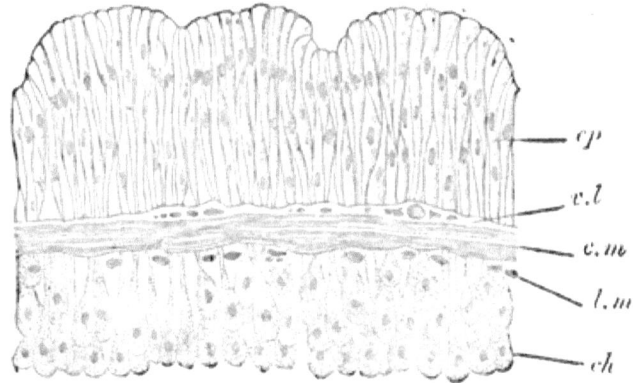

Fig. 40.—Highly magnified cross-section through the wall of the alimentary canal. *ch*, chloragogue layer; *c.m*, circular muscles; *e.p*, lining epithelium; *l.m*, longitudinal muscles; *v.l*, vascular layer.

Blood-vessels appear in the section as rounded or irregular cavities bounded by thin walls. They consist of a delicate lining epithelium covered by a thin layer of muscle-fibres. In the walls of the stomach-intestine the vessels are often completely invested by chloragogue-cells, which radiate from them with great regularity (Fig. 39). The finer branches have no muscular layer, consisting of the epithelium alone.

Dissepiments. These often appear in cross or longitudinal sections. They consist chiefly of muscle-fibres irregularly disposed, intermingled with connective-tissue cells and fibres, and covered on both sides with the peritoneal epithelium.

Nervous System. A cross-section of a ganglion (Fig. 41) shows it to be composed of two distinct parts, viz., (1) the gan-

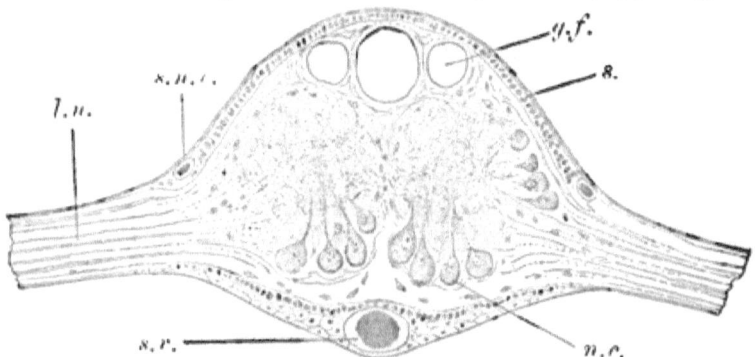

FIG. 41.—Highly magnified cross-section of a ventral ganglion. *g.f*, giant-fibres; *l.n*, lateral nerve; *n.c*, nerve-cells; *s*, muscular sheath of the ganglion; *s.v*, sub-neural vessel; *s.n.v*, supra-neural vessel.

glion proper on the inside, and (2) a sheath which envelops it. The sheath (*s*, Fig. 41) consists of two layers, viz. :—

1. *Peritoneal Epithelium.* On the outside.
2. *Muscular Layer*, or sheath, a thick layer of irregularly arranged muscle-fibres intermingled with connective tissue. Imbedded in it are the sub-neural blood-vessel on the lower side and the supra-neural blood-vessels on each side above. In the middle line are three rounded spaces (*g, f.* Fig. 41), which are the cross-sections of three hollow fibres running along the entire length of the ventral nerve-chain. They are called "giant-fibres," and possibly serve to support the soft parts of the nerve-cord.

The *Ganglion* proper is distinctly bilobed, and consists of two portions, viz. :—

1. *Nerve-cells* (*n.c*). Numerous pear-shaped nerve-cells near the surface, with their narrow ends turned towards the centre, into which each sends a single branch or nerve-fibre. They are confined chiefly to the ventral and lateral parts of the ganglion.

2. *Fibrous Portion.* This occupies the central part. It consists of a close and complicated network of nerve-fibres intermingled with connective tissue. Some of these fibres communicate with branches of the nerve-cells, as stated above; others run out into the lateral nerves, while still others run along the commissures to connect with fibres from other ganglia.

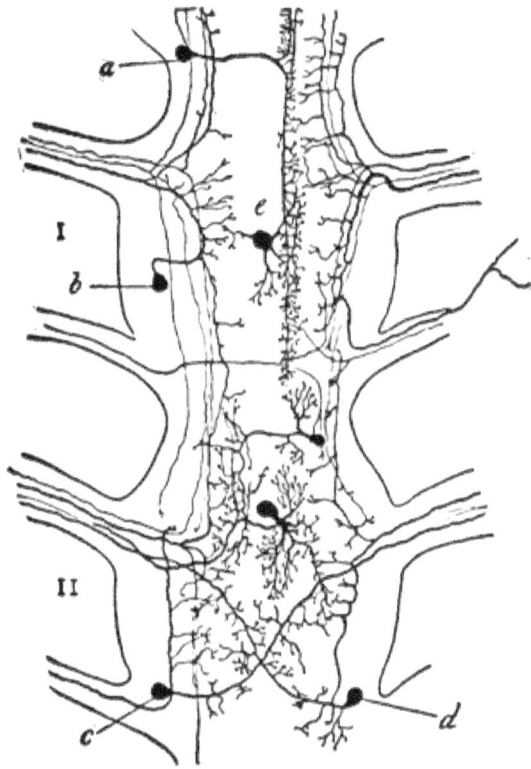

FIG. 42.—Two of the ventral ganglia (I, II) of *Lumbricus* with the lateral nerves, showing some of the motor nerve-cells and fibres (black). *a* sends fibres forwards and backwards within the nerve-cord; *b*, a fibre into one of the double nerves on its own side; *c* and *d*, fibres that cross to the nerves of the opposite side. (After Retzius.)

According to the latest researches (of Lenhossék and Retzius) most if not all of the nerve-cells of the ventral cord are motor in function. Near the centre of each ganglion (Fig. 42, *e*) in a single large multipolar cell of doubtful nature. All the other cells are either bipolar or unipolar, in the latter case sending out a single branch which soon divides into two. In every case one of the branches breaks up into fine sub-divisions within the cord. The other branch in most cases passes out of the cord through one of the lateral nerves to the muscles or other peripheral organs, either crossing within the cord to the opposite side of the body or making exit on its own side. Some of the cells, however, are purely "commissural," i.e., neither branch leaves the cord.

The sensory fibres entering from the periphery terminate freely (not in nerve-cells), breaking up into numerous fine branches on the same side of the cord. (Fig. 43.)

The nerves leaving the central system are mixed, i.e., they contain both sensory and motor fibres.

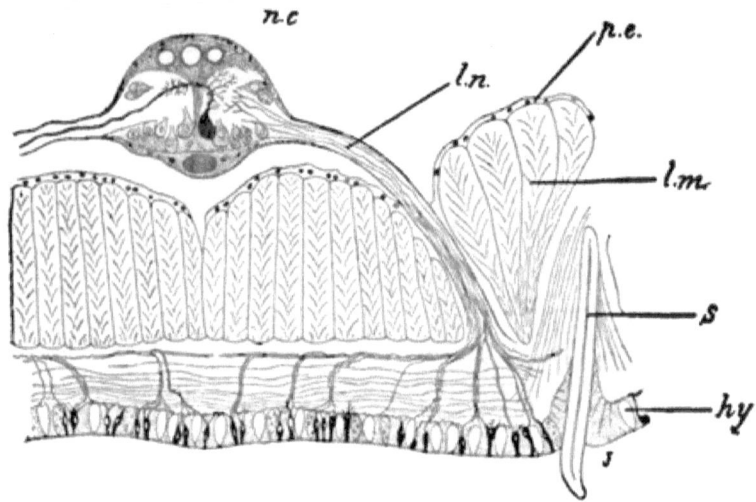

Fig. 43.—Transverse section of ventral part of the body, showing the nervous connections. *n.c*, ventral ganglion, giving off a lateral nerve at *l.n.*; *p.e.*, peritoneal epithelium; *l.m.*, longitudinal muscles; *hy*, hypodermis; *s*, seta. A single motor nerve-cell (black) is shown sending a fibre into the nerve towards the left. In the nerve to the right are sensory fibres proceeding inward from the sensory cells (black) of the hypodermis, and terminating in branching extremities. (After Lenhossék.)

Sections through the ventral commissures are similar to those through the ganglia, but the central portion (i.e., that within the sheath) is smaller, is divided into two distinct parts, and the nerve-cells are less abundant.

Sections through the nerves show them to consist only of parallel fibres surrounded by a sheath which gradually fades away as the nerves grow smaller, and finally disappears, the muscular layer first disappearing, and then the epithelial covering.

With this brief sketch of the histological structure of the earthworm we conclude our morphological study of the animal. Those who desire fuller information on the histology will find a general treatment of it in the work of Claparède, already cited at p. 93. Many later works have been published on the detailed histology.

CHAPTER VII.

THE BIOLOGY OF AN ANIMAL (*Continued.*)

Physiology of the Earthworm.

In the preceding pages brief descriptions of many special physiological phenomena have been given in connection with the detailed descriptions of the primary functions and systems. It now remains to consider the more general problems of the life of the animal, and especially its relations to the environment, and the transformations of matter and energy which it effects.

The Earthworm and its Environment. The earthworm is an organized mass of living matter occupying a definite position in space and time, and existing amid certain definite and characteristic physical surroundings which constitute its "environment."

As ordinarily understood the term environment applies only to the immediate surroundings of the animal—to the earth through which it burrows, the air and moisture that bathe its surface, and the like. Strictly speaking, however, the environment includes everything that may in any manner act upon the organism—that is, the whole universe outside the worm. For the animal is directly and profoundly affected by rays of light and heat that travel to it from the sun; it is extremely sensitive to the alternations of day and night, and the seasons of the year; it is acted on by gravity; and to all these, as well as to more immediate influences, the animal makes definite responses.

We have seen that the body of the earthworm is a complicated piece of mechanism constructed to perform certain definite actions. But every one of these actions is in one way or another dependent upon the environment and directly or indirectly relates to it. At every moment of its existence the organism is acted on by its environment; at every moment it reacts upon the environment, maintaining with it a constantly shifting state of equilibrium which finally gives way only when the life of the animal draws to a close.

Adaptation of the Organism to its Environment. In its relations to the environment the earthworm embodies a fundamental biological law, viz., that *the living organism must be adapted to its environment*, or, in other words, that a certain *harmony* between organism and environment is essential to the continuance of life, and any influence which tends to disturb or destroy this harmony tends to disturb or destroy life. The adaptation may be either passive (structural) or active (functional). Structural adaptation is well illustrated, for instance, by the general shape of the body, so well adapted for burrowing through the earth. Again, the delicate integument gives to the body the flexibility demanded by the peculiar mode of locomotion; it affords at the same time a highly favorable respiratory surface—a matter of no small importance to the worm in its badly-ventilated burrow; and yet this delicate integument does not lead to desiccation, because the animal lives always in contact with moist earth. The alimentary canal, long and complicated, is most perfectly fitted for working over and extracting nutriment from the earthy diet. The reproductive organs are a remarkable instance of complex structural adaptation in an animal which on the whole is of comparatively simple structure.

Functional adaptation is perhaps best shown in the instinctive actions or "habits" of the worm. Its nocturnal mode of life (functional adaptation to light) and its "timidity" protect it from heat, desiccation, from birds and other enemies. In winter or in seasons of drought it burrows deep into the earth.

A striking instance of adaptation is shown in the care which is taken to insure the welfare of the embryo worms. Minute, delicate, and helpless as they are, they develop in safety inside the tough, leathery capsule (p. 78), floating in a milklike liquid which is at once their cradle and their food.

Origin of Adaptations. The development of the earthworm shows that its whole complex bodily mechanism takes origin in a single cell (p. 74), and that all the remarkable adaptations expressed in its structure and action are brought about by a *gradual process* in the life-history of each individual worm. There is reason to believe that this is typical of the ancestral history (descent) of the species as a whole, and that adaptation has been gradually acquired in the past. We know that environments change, and that to a certain extent organisms change correspondingly through functional adaptation, provided the change of environment be not too sudden or extreme. In other words, the organism possesses a certain *plasticity* which enables it to adapt itself to gradually-changing conditions of the environment.

Now there is good reason to believe that as environment has gradually undergone changes in the past, organisms have gradually undergone corresponding changes of structure. Those which have become in any way so modified as to be most perfectly adapted to the changed environment have tended to survive and leave similarly-adapted descendants. Those which have been less perfectly adapted have tended to die out through lack of fitness for the environment; and by this process—called by Darwin "Natural Selection" and by Spencer the "Survival of the Fittest"—the remarkable adaptations everywhere met with are believed to have been gradually worked out.

It should be observed that Natural Selection does not really explain the *origin* of adaptations, but only their persistence and accumulation. The theory of evolution is not at present such as to enable us to say with certainty what causes the first origin of adaptive variations.

Nutrition. The earthworm does work. It works in travelling about and in forcing its way through the soil; in seizing, swallowing, digesting, and absorbing food; in pumping the blood; in maintaining the action of cilia; in receiving and sending out nerve-impulses; in growing; in reproducing itself—in short, in carrying on any and every form of vital action. To live is to work. Now work involves the expenditure of energy, and the animal body, like any other machine, while life continues, requires a continual supply of energy. It is clear from what has been said on p. 32 that the immediate source of the energy expended in vital action is the working protoplasm itself, which undergoes a destructive chemical change (katabolism or destructive metabolism) having the nature of an oxidation. From this it follows on the one hand that the waste products of this action must be ultimately passed out of the body as excretions, and on the other hand that the loss must ultimately be made good by fresh supplies entering the animal in the form of food. It is further evident that the income must equal the outgo if the animal is merely to hold its own, and must exceed it if the animal is to grow.

Thus it comes about that there is a more or less steady flow of matter and of energy through the living organism, which is itself a centre of activity, like a whirlpool (p. 2). The chemical phenomena accompanying the flow of matter and energy through the organism are those of *nutrition* in the widest sense. This term is more often restricted especially to the phenomena accompanying the income, while those pertaining to the outgo are regarded as belonging to *excretion*. The intermediate processes directly connected with the life of protoplasm are put together under the head of *metabolism*; they include both the constructive processes by which protoplasm is built up (*anabolism*) and the destructive processes by which it is broken down (*katabolism*) in the liberation of energy.

Income. It is difficult to determine the exact income of *Lumbricus*, but it may be set down approximately as follows:—

INCOME OF *LUMBRICUS*.

MATTER.	WHENCE DERIVED.
1. *Proteids.*	From vegetal or animal matters taken in through the mouth.
2. *Fats.*	From vegetal or animal matters taken in through the mouth.
3. *Carbohydrates.*	From vegetal or animal matters taken in through the mouth.
4. *Water.*	Taken in through the mouth, or perhaps to some extent absorbed through the body-walls.
5. *Free oxygen.*	Absorbed directly from the atmosphere or ground-air by diffusion through the body-walls. Sometimes from water in which it is dissolved.
6. *Salts.*	Various inorganic salts taken along with other food-stuffs.
ENERGY.	
Potential.	In the food.

The food-stuffs are converted by the animal into the substance of its own body (protoplasm and all its derivatives), and they must therefore be the ultimate source of energy. It follows that the animal takes in energy only in the potential form (i.e., in the chemical potential between the oxidizable proteids, carbohydrates and fats, and free oxygen). It is true that the animal may under certain circumstances absorb kinetic energy in the form of heat, but this is available only as a *condition*, not as a cause of protoplasmic action. In this inability to use kinetic energy the earthworm is typical of animals as a whole.

Of the organic portion of the food proteids are a *sine qua non*, and in this respect again the worm is a type of animal life in general. Either the fats or the carbohydrates may be omitted (though the animal probably thrives best upon a mixed diet in which both are present), but without proteids no animal, as far as is known, can long exist.

General History of the Food. Digestion and Absorption.
Lumbricus takes daily into its alimentary canal a certain amount of necessary food-stuffs, but these are not really inside the body so long as they remain in the alimentary canal; for this is shown by its development to be only a part of the outer surface folded in to afford a safe receptacle within which the food may be worked over. Before the food can be actually taken into the body, or *absorbed*, it must undergo certain chemical changes collectively called *digestion* (cf. p. 49). A very important part of this process consists in rendering non-diffusible substances diffusible, in order that they may pass through the walls of the alimentary canal into the blood. Proteids, for example, have been shown to be non-diffusible (Chap. III). In digestion they are changed by the fluids of the alimentary canal into *peptones* —substances much like proteids, but readily diffusible. In like manner the non-diffusible starch is changed into diffusible sugar and becomes capable of absorption. It is highly probable that all carbohydrates are thus turned into sugar. The fats are probably converted in part into soluble and diffusible soaps which are readily absorbed, but are mainly emulsified and directly passed into the cells of the alimentary tract in a finely divided state. Nothing, however, is known of this save by analogy with higher animals. In all cases digestion takes place *outside the body*, and is only preliminary to the real entrance of food into the physiological, or true, interior.

Metabolism. After absorption into the body proper the incoming matters are distributed by the circulation to the ultimate living units or cells, and are finally taken up by them and built into their substance. There is reason to believe that each cell takes from the common carrier, the blood, only such materials as it needs, leading a somewhat independent life as to its own nutrition. It co-operates with other cells under the direction of the nervous system (co-ordinating mechanism), but to a great degree is independent in its choice of food—just as a soldier in a well-fed army obeys orders for the common good, but yet takes only what he chooses from the daily ration supplied to all.

What takes place within the cell upon the entrance of the food is almost wholly unknown, but somehow the food-matters, rich in potential energy, are built up into the living substance probably by a series of constructive processes culminating in protoplasm. Alongside these constructive processes (anabolism) a continual destructive action goes on (katabolism); for living matter is decomposed and energy set free in every vital action, and vitality or life is a continuous process. It must not be supposed, however, that either the synthetic or the destructive process is a single act. Both probably involve long and complicated chemical transformations but the precise nature of these changes is at present almost wholly unknown. It is certain that the destructive action is in a general way a process of oxidation effected by aid of the free oxygen taken in in respiration. We may be sure, however, that it is not a case of simple combustion (i.e., the protoplasm is not "burnt"). It is more probably analogous to an explosive action, the oxygen first entering into a loose association with complex organic substances in the protoplasm, and then suddenly combining with them under the appropriate stimulus to form simpler and more highly-oxidized products. Of the precise nature of the process we are quite ignorant.

Outgo. Just as the income of the animal represents only the first term in a series of constructive processes, so the outgo is the last term of a series of destructive actions of which we really know very little save through their results. The outgo is shown in the accompanying table.

Both energy and matter leave the cells, and finally leave the body—the former as heat, work done, or energy still potential (in urea and other organic matters); the latter as excretions, which diffuse freely outwards through the skin and nephridial surfaces.

OUTGO OF *LUMBRICUS*.

MATTER.	MANNER OF EXIT.
Carbon dioxide (CO_2).	Mainly by diffusion through the skin.
Water (H_2O).	Through the skin, through the nephridia, and in the fæces.
Urea [$(NH_2)_2CO$], and its allies.	Through the nephridia.
Salts.	Dissolved in the water.
Proteids and other organic matters.	In the substance of the germ-cells, the egg-capsules, and the contained nutrient fluids.
ENERGY.	
Potential.	A small amount still remaining in urea, in the germ-cells, etc.
Kinetic.	Work performed. Heat.

Of the daily outgo the water, carbon dioxide, and salts are devoid of energy, but the urea contains a small amount which is a sheer loss to the animal. Were the earthworm a perfect machine it could use this residue of energy by decomposing the urea into simpler compounds [viz., ammonia (NH_3), carbon dioxide (CO_2), and water (H_2O)]; but it lacks this power, though there are certain organisms (*Bacteria*) which are able to utilize the last traces of energy in urea (p. 197). To the daily outgo must be added the occasional loss both of matter and of energy suffered in giving rise to ova and spermatozoa, and in providing a certain amount of food and protection for the next generation.

Interaction of the Animal and the Environment. The action of the environment upon the animal has already been sufficiently stated (p. 97). It remains to point out the changes worked by the animal on the environment. These changes are of two kinds, mechanical (or physical) and chemical. The most important of the former is the continual transformation of the soil which the worms effect, as Darwin showed, by bringing the deeper layers to the surface, where they are exposed to the atmosphere, and also by dragging superficial objects into the burrows. The chemical changes are still more significant. The general effect of the metabolism of the animal is the destruction by oxidation of organic matter; that is, matter originally taken from the environment in the form of complex proteids, fats, and carbohydrates is returned to it in the form of simpler and more highly oxidized substances, of which the most important are carbon dioxide and water (both inorganic substances). This action furthermore is accompanied by a dissipation of energy—that is, a conversion of potential into kinetic energy.

On the whole, therefore, the action of the animal upon the environment is that of an oxidizing agent, a reducer of complex compounds to simpler ones, and a dissipator of energy. And herein it is typical of animals in general.

CHAPTER VIII.

THE BIOLOGY OF A PLANT.

The Common Brake or Fern.

(*Pteris aquilina*, Linnæus.)

For the study of a representative vegetal organism some plant should be chosen which may be readily procured and is neither very high nor very low in the scale of organization. Such a plant is a common fern.

Ferns grow generally in damp and shady places, though they are by no means confined to such localities. Some of the more hardy species prefer dry rocks or even bold cliffs, in the crevices of which they find support; others live in open fields or forests, and still others on sandy hillsides. In the northern United States there are altogether some fifty species of wild ferns, but those which are common in any particular locality are seldom more than a score in number. Throughout the whole world some four thousand species of ferns are known, but by far the greater number are found only in tropical regions, where the climate is best suited to their wants. At an earlier period of the earth's history ferns attained a great size, and formed a conspicuous and important feature of the vegetation. At present, however, they are for the most part only a few feet in height. Nearly all are perennial; that is, they may live for an indefinite number of years. Most of them have creeping or subterranean stems; but some of the tropical species have erect, aërial stems, sometimes rising to a height of fifty feet or more and forming a trunk which is cylindrical, of equal diameter throughout, and bears leaves only at the summit, like a palm (tree-ferns).

Of all the ferns perhaps the commonest and most widely distributed is the "brake" or "eagle-fern," which is known to botanists as *Pteris aquilina*, Linnæus, or *Pteridium aquilinum*,

Kuhn. This plant is not only common, but of comparatively simple structure; it is of a convenient size, and has been much studied. It may therefore be taken both as a representative fern and as a representative of all higher vegetal organisms.

Habitat, Name, etc. The brake occurs widely distributed in the United States, under a great variety of conditions; e.g., in loose pine groves, especially in sandy regions; in open woodlands amongst the other undergrowth; on hillside pastures and in thickets—indeed almost everywhere, except in very wet or very dry places. It appears to be equally common elsewhere; for, according to Sir W. J. Hooker, *Pteris aquilina* grows " all round the world, both within the tropics and in the north and south temperate zones. . . . In Lapland it just passes within the Arctic circle, ascending in Scotland to 2000 feet, in the Cameroon Mountains to 7000 feet, in Abyssinia to 8000 or 9000 feet, in the Himalayas to about 8000 feet." (*Synopsis Filicum.*)

" Pteris ($\pi\tau\epsilon\rho\iota\varsigma$, the common Greek name for *fern*), signifying wing or feather, well accords with the appearance of *Pteris aquilina*, the most common and most generally distributed of European ferns. It is possible that this fern may rank as the most universally distributed of all vegetable productions, extending its dominion from west to east over continents and islands in a zone reaching from Northern Europe and Siberia to New Zealand, where it is represented by, and perhaps identical with, the well-known *Pteris esculenta*. The rhizome of our plant like that of the latter is edible, and though not employed in Great Britain as food, powdered and mixed with a small quantity of barley-meal it is made into a kind of gruel called *gofio*, in use among the poorer inhabitants of the Canary Islands."— (Sowerby.)

The specific name *aquilina* (*aquila*, eagle) and a popular name, "eagle-fern," in Germany, etc., have come from a fanciful likeness of the dark tissue seen in a transverse section of the leaf-stalk to the figure of an outspread eagle. The same figure has, however, been compared to an oak-tree, and has also given rise to the name of " devil's-foot fern," from its alleged resemblance to " the impression of the deil's foot," etc., etc.

The popular designation of this plant as " the brake " testifies to its great abundance; for a brake is a dense thicket or undergrowth—as for example a cane "brake."

When fully grown (Fig. 44) the common brake has a leafy top supported by a polished, dark-colored, erect stem, which in New England rises to a height of from one to four feet above the ground. In this climate, however, it appears to be somewhat undersized, for it grows to a height of fourteen feet in the Andes,* and in Australia attains to twice the height of a man, forming a dense undergrowth beneath tree-ferns 40–100 feet high.† In Great Britain it is from six inches to nine feet high (Sowerby), or even larger in exceptional cases. "In dry gravel it is usually present, but of small size; while in thick shady woods having a moist and rich soil it attains an enormous size, and may often be seen climbing up, as it were, among the lower branches and underwood, resting its delicate pinnules on the little twigs, and hanging gracefully over them." (Newman.)

GENERAL MORPHOLOGY OF THE BODY.

The body of the fern, like that of the earthworm, consists of cells, grouped to form tissues and organs. Their arrangement, however, differs widely from that in the animal, for the plant-body is a nearly solid mass, and there are no extended internal cavities enclosing internal organs. The organs of the plant are for the most part external, and arise by local modifications of the general mass. Like many higher plants the body of the fern consists of an *axis* or stem-bearing branches, from which arise leaves. The fern differs form ordinary trees, however, in the fact that the stem, with its branches, lies horizontal beneath the surface of the ground. Only the leaves (fronds) rise into the air. (Fig. 44.) It is convenient to describe the body of the brake, accordingly, as consisting of two very different parts—one green and leaflike, which rises above the ground; the other black and rootlike, lying buried in the soil. These will henceforth be spoken of as the *aerial* and the *underground* parts.

The *underground part* lies at a depth of an inch to a foot

---

* Hooker, *l. c.*
† Krone, *Botan. Jahresbericht,* 1876 (4), 346.

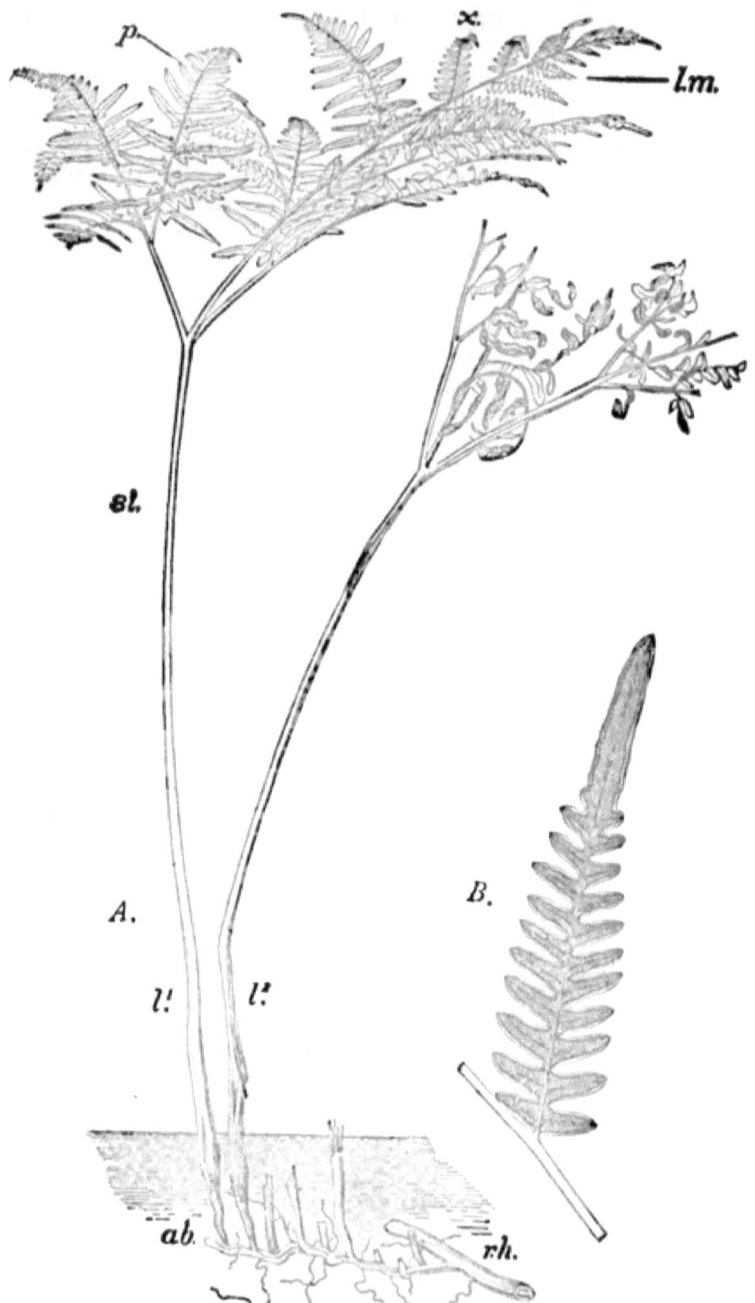

FIG. 44.—The Brake (*Pteris aquilina*), showing part of the underground stem (*r.h*) and two leaves, one (*l¹*), of the present year, in full development; the other (*l²*), of the past year, dead and withered. *a.b*, apical bud at the extremity of a branch which bears the stumps of leaves of preceding years and numerous roots; *l¹*, mature active leaf; *l²*, dead leaf of preceding year; *l.m*, lamina of leaf; *p*, pinna; *r.h*, portion of main rhizome; *x*, younger pinna, which is shown enlarged at *B*. This pinna is nearly similar to the pinnules of older pinnæ. (× ½.)

below the surface, and branches widely in various directions. It may often be followed for a long distance, and in such cases reveals a surprisingly complicated system of underground branches. At first sight, the underground portion of the fern appears to be the root, but a closer examination shows it to be really the *stem* or *axis* of the plant, which differs from ordinary stems chiefly in the fact that it lies horizontally under the ground instead of rising vertically above it. The aerial portion, which is often taken for stem and leaf, is really leaf only. The true roots are the fine fibres which spring in great abundance from the underground stem. Underground stems more or less like that of *Pteris* are not uncommon—occurring, for instance, in the potato, the Solomon's-seal, the onion, etc. In *Pteris*, and in certain other cases, the underground stem is technically called the *rootstock* or *rhizome*, and in this plant it constitutes the larger and more persistent part of the organism. In the specimen shown in Fig. 45 the rhizome was about eight feet long and bore two leaves. It was dug out of sandy soil on the edge of a woodland, and lay from one to six inches below the surface. It was crossed and recrossed in all directions, both above and below, by the rhizomes of its neighbors, the whole constituting a coarse network of underground stems loosely filling the upper layer of the soil.

The *aerial portion* (the *frond* or *leaf*) is likewise divisible into a number of parts, comprising in the first place the leaf-stalk or *stipe*, and the leaf proper or *lamina*. The latter is subdivided like a feather (*pinnately*) into a number of lobes (*pinna*, Fig. 44), which vary in form according to the state of development of the leaf. In large leaves the two lower pinnæ are often larger than the others, so that the leaf appears to consist of three principal divisions, and is said to be "*ternate*" or triply divided (Fig. 44, *A*). Each pinna is in turn pinnately subdivided into *pinnules* (*pinnula*) or leaflets (Fig. 44, *B*), each of which is traversed down the middle by a thickened ridge or rod, the *midrib*. The leaflets sometimes have smooth outlines, but are usually lobed along the edges, as in Fig. 44, *B*. In this case their form is said to be *pinnatifid*. Each lobe is likewise furnished with a midrib. The *stipe* enlarges somewhat just below the surface of the ground, then grows smaller and

Fig. 45.—An entire plant of *Pteris*. One of the leaves is young and small, and a comparison of the figure with Fig. 44 will show some of the differences between leaves of different ages.

joins the rhizome. The enlargement is of considerable interest, for it occurs at precisely the point of greatest strain when the leaf is bent by the wind or otherwise, and must serve to strengthen the stipe.

It will appear from the following description that the plant body exhibits in some measure certain general forms of symmetry and differentiation which in a broad sense may be regarded as analogous to those occurring in the animal. The rhizome grows only at one end, and in its structure suggests the antero-posterior differentiation of the animal. It also shows a slight differentiation between the upper and lower surfaces, which appears both in the external form and in the arrangement of the internal lines. It is furthermore distinctly bilateral, a vertical plane dividing it into closely similar halves. These features are, however, far less prominent in the fern than in the earthworm, and in plants they never attain a high degree of development, while in the higher animals they are among the most conspicuous and important features of the body. Of more general importance in the fern is the repetition of similar parts (branches, roots, leaves) along the axis, which suggests, perhaps, a certain analogy to animal metamerism, though not usually recognized or designated by the same term. All of these conditions of differentiation and symmetry are more easily made out by an examination of the aerial portion.

The plant as a whole, may be regarded as consisting of an *axis* (the rhizome and its branches) which bears a number of *appendages* in the form of roots and leaves. The axis forms the central body or trunk of the plant, and in it most of its matter and energy are stored; the appendages are organs for taking in food, for excretion, for respiration, for reproduction, etc.

The Underground Stem, or Rhizome, and its Branches. The *rhizome* is a hard black, elongated, and branching stem, generally flattened somewhat in the vertical direction as it lies in the earth, and expanded slightly on either side to form well-marked lateral folds—the *lateral ridges*. Its thickness is seldom more than half an inch, and usually considerably less. In transverse section it has the outline shown in Fig. 48, and the marginal part only is black. The branches repeat in all respects the form and structure of the main axis. Both the main axis and the branches end either in conical, pointed, and fleshy structures about two inches long, or in blunt, yellowish knobs, plainly depressed in the centre. At these ends the rhizome grows; hence they are called the growing points or *apical buds* (Figs. 44, 47).

Besides the apical buds the rhizome bears nearly always one or more dead, decaying tips. These arise in the following manner: After attaining a certain length both the rhizome and its branches gradually die away behind. Death of the hinder part follows at about the same rate with which growth advances at the apical buds; so that the total length may not change materially from year to year. It is obvious that this process must result in the gradual and successive detachment of the branches from the main axis. Each branch, now become an independent rhizome, repeats the process; and in this manner a single original rhizome may give rise to large numbers of distinct plants, all of which have been at some time in material connection with an ancestral stock. This process is evidently a kind of *reproduction* (though it is not the most important or most obvious means for the propagation of the plant), and in this way a large area may be occupied by distinct, though related, plants whose branching rhizomes cross and recross, making the subterranean network already described, p. 109.

Origin of Leaves upon the Rhizome and its Branches. The young plant of *Pteris* puts up a number of leaves (7-12) yearly, but the adult generally develops one only, which grows very slowly, requiring two years before it unfolds. Towards the end of the first year it is recognizable only as a minute knob at the bottom of a depression near the growing point. At the beginning of the second year it is perhaps an inch high, the stalk

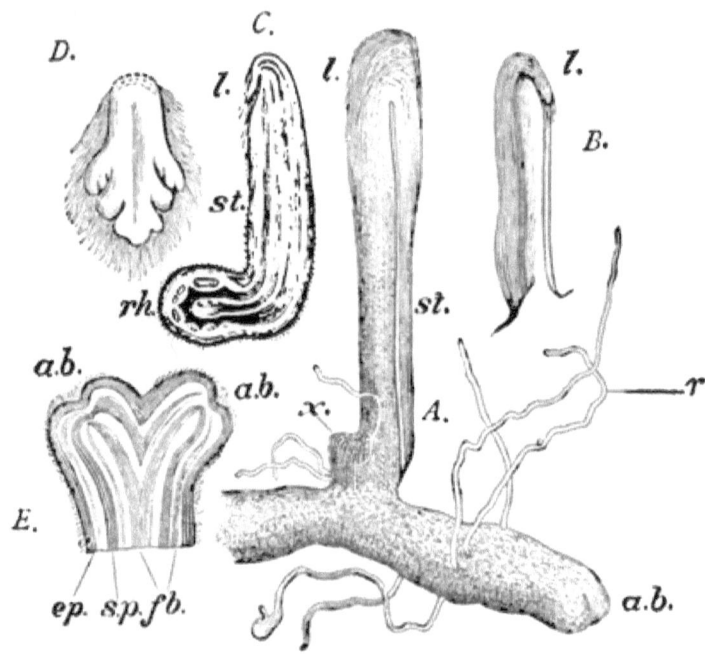

FIG. 46. (After Sachs.)—Developing leaf, etc., of *Pteris*. *A*, end of a branch showing the apical bud and the rudiment of a leaf; *B*, a rudimentary leaf; *C*, a similar leaf in longitudinal section, showing the infolded lamina (*l*), the attachment to the rhizome, and the prolongation of the tissues of the latter into the leaf; *D*, lamina of a very young leaf; *E*, horizontal section through a growing point which has just forked to form two apical buds. *a.b.* apical bud; *ep*, epidermis and underlying sclerotic parenchyma; *f.b*, fibro-vascular bundles; *l*, lamina; *r*, root; *s.p*, sclerotic prosenchyma; *x*, an adventitious bud at the base of the leaf.

only having appeared. At the end of the second year the lamina is developed, and hangs down as shown in Fig. 46, *C*. Early in the spring of the third year it breaks through the ground, and grows rapidly to the fully-matured state.

LEAVES AND RHIZOME. 113

The leaves usually arise near the apical buds of the main axis or of the branches. Behind each mature leaf remnants of the leaves of preceding years are often to be found, alternating on the sides of the rhizome in regular succession, and showing various stages of decay. The first of these (which is on the opposite side of the rhizome from the living leaf) was alive the previous year; the next (on the same side with the living leaf) is the leaf of the year before that; and so on. Fig. 47 shows an example of this sort. The leaf of the present year, l, is fully

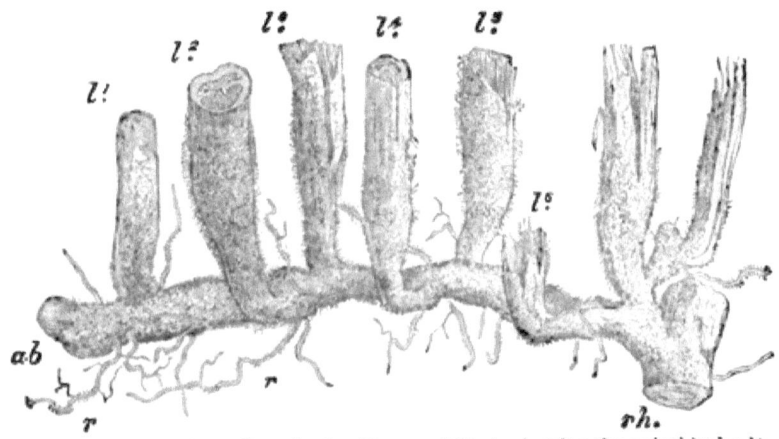

FIG. 47. (After Sachs.)—Branch of a rhizome of *Pteris*, showing the apical bud (a,b), the stumps of a number of successive leaves (l^2, l^3, l^4, etc.), and a part of the main rhizome (rh), r, root.

developed; and the relics of the leaves of the preceding years are indicated at l^2, l^3, etc.; l^1 is the rudiment of next year's leaf.

Internal Structure of the Rhizome. The rhizome is a nearly solid mass, consisting of many different kinds of cells, united into different tissues, and having a very complicated arrangement. Its study is somewhat difficult. Nevertheless the arrangement of the cells is definite and constant, and merits careful attention, since it has many features which are characteristic of the cellular structure of the stems of higher plants. We shall first examine its more obvious anatomy as displayed in transverse and longitudinal sections, afterwards making a careful microscopical study of the cells and tissues.

Seen with a hand-lens or the naked eye, a transverse section of the rhizome (Fig. 48) presents a white or yellowish background bounded by a black margin (the *epidermis*) and marked by various colored or pale spots and bands; the latter are different tissues, or systems of tissue. These different structures are arranged in three groups or *systems of tissue*, which are found

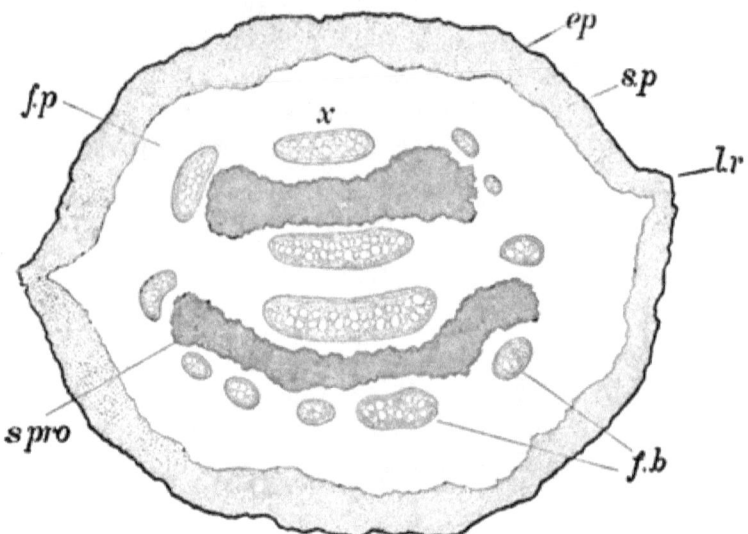

FIG. 48. Cross-section of the rhizome of *Pteris*. *l.r*, lateral ridges; *f.p*, fundamental parenchyma; *s.p*, sclerotic parenchyma; *s.pro*, sclerotic prosenchyma; *f.b*, *x*, fibro-vascular bundles.

among all higher plants in essentially the same form, though differing widely in the minor details of their arrangement. These are:—

   I. The Fundamental System of Tissues.
   II. The Epidermal System.
   III. The Fibro-vascular System.

The *Fundamental system* consists in *Pteris* of three tissues:
(*a*) *fundamental parenchyma* (Fig. 48, *f.p*), the soft whitish mass forming the principal substance of the rhizome;

(*b*) *sclerotic parenchyma* (*s.p*), the brown hard tissue lying just below the epidermis, from which it is scarcely distinguishable;

(*c*) *sclerotic prosenchyma* (*s.pro*), black or reddish dots and bands of extremely hard tissue, most of which is contained in two conspicuous bands lying one on either side of a plane joining the lateral ridges.

The sclerotic parenchyma and the sclerotic prosenchyma both arise through a transformation (hardening, etc.) of portions of originally-soft fundamental parenchyma. In most plants above the ferns the fundamental system contains neither of these tissues.

The *Fibro-vascular system* is composed of longitudinal threads or strands of tissue known as the *fibro-vascular bundles*, and these in one form or another are characteristic of all higher plants. They appear here and there in the section (Fig. 48, *f.b*) as indistinct, pale or silvery areas of a roundish, oval, or elongated shape. Closely examined they show an open texture, enclosing spaces which are sections of empty tubes, or vessels and fibres, from which the bundles take their name.

The *Epidermal system* consists of a single tissue, the *epidermis*, which covers the outside of the rhizome.

By a simple dissection of the stem with a knife the sclerotic prosenchyma and the fibro-vascular bundles may be seen to be long strands or bands, coursing through the softer fundamental tissues.

It should be clearly understood that these three systems are, in general, not single tissues, but *groups* of tissues which are constantly associated together for the performance of certain functions.*

Microscopic Anatomy (Histology) of the Rhizome.

General Account. Microscopic study of thin sections of the rhizome shows the various tissues to be composed of innumerable closely-crowded cells, which differ very widely in structure and in function. In studying these cells the student should not lose sight of the fact that they are objects having three dimensions, of which only two are seen in sections. And hence a single section may give an imperfect or entirely false impression of the real form of the cells,—just as the face of a wall of masonry may give only an imperfect idea of the blocks of which it is built.

---

* This classification of the tissues is only a matter of convenience, and has little scientific value. By many botanists it has been rejected altogether; but no apology for its use need be made by those who, like the authors, have found it useful, so long as it is defended by Sachs (who first introduced it) and its value for beginners is conceded by De Bary.

For this reason many of the cells can only be understood by a comparison of transverse and longitudinal sections, and these should be studied together until their relations are thoroughly mastered.

The following table gives brief definitions of the leading vegetal tissues and is good not only for *Pteris* but for all plants :—

PRINCIPAL ADULT VEGETAL TISSUES.

TISSUES.	CHARACTERISTICS.
1. *Epidermis.*	Cells in a single layer covering the outer surface.
2. *Parenchyma.*	Masses of cells, rounded, prismatic or polyhedral, usually incompletely joined at the angles, thus leaving intercellular spaces. Not much longer than broad. Thin-walled.
3. *Prosenchyma.*	Cells elongated, typically massed, without intercellular spaces.
4. *Sieve-tubes.*	Cells elongated, thin-walled, panelled with perforated areas, containing proteids.
5. *Tracheids.*	Cells thick-walled, elongated, pointed, hard; walls pitted; filled with air.
6. *Trachea* or *vessels.*	Cells very slender, elongated, opening into one another at their ends, often spirally thickened, and filled with air.

These six tissues are not only found in the rhizome, but extend throughout the roots and the fronds as well. Moreover, all the tissues not only of the fern but of all higher plants are varieties of them.

Special Account. It must not be forgotten that the differences between tissues are only the outcome of the differences between their component cells (p. 13). So that the study of the histology of the rhizome, even if preceded (as it may well be) by a dissection, and a naked-eye examination of some of the tissues, eventually resolves itself into the careful microscopic study of the several kinds of cells composing those tissues.

The mature parts of the rhizome contain at least nine very different kinds of cells, the characteristics and grouping of which are shown in the following table. In the apical buds, however, this arrangement disappears, and all the cells appear closely similar.

MINUTE ANATOMY OF THE RHIZOME OF *PTERIS AQUILINA*.

System.	Tissues.	Characteristics.
I. Epidermal	1. *Epidermis.*	Cells polygonal in cross-section, empty. Walls hard, thickened, especially towards the outside. (Fig. 49.)
II. Fundamental.	2. *Fundamental parenchyma.*	Cells rounded or polygonal in cross-section, colorless. Thin-walled, containing protoplasm, nucleus and starch. Intercellular spaces present. (Fig. 52, *f.p.*)
	3. *Sclerotic parenchyma.*	Cells polygonal or semi-fusiform in section, nearly empty. No intercellular spaces. Walls hard and brown, thickened. (Fig. 49.)
	4. *Sclerotic prosenchyma* (or *sclerenchyma*)	Cells fusiform, empty. Walls thick, red. (Fig. 50.)
III. Fibro-vascular.	5. *Wood-parenchyma.*	Like the fundamental parenchyma, but with more elongated cells. (Figs. 52, 53.)
	6. *Phloëm-parenchyma.*	Precisely like 5, differing only in position.
	7. *Phloëm-prosenchyma, or bast-fibres.*	Cells fusiform, rich in protoplasm, colorless. Walls thick, soft. (Figs. 52, 53.)
	8. *Sieve-tubes.*	Having the ordinary characters (see preceding table). (Figs. 52, 54.)
	9. *Tracheïdes* (or *ladder-cells*).	Pits transversely elongated (scalariform). (Figs. 52, 53.)
	10. *Tracheæ or vessels (spiral).*	Very slender, with one or two internal spiral thickenings. (Fig. 52.)

Besides the above-mentioned tissues, the rhizome contains certain other secondary varieties which will be described further on.

Epidermal System. Epidermis. It is the function of the epidermis (aided in this case by the underlying sclerotic parenchyma) to protect the inner tissues from contact with the soil and to guard against desiccation of the rhizome during droughts. The cells (Fig. 49) are dead and empty, with enormously thick, hard walls perforated by numerous branching canals. The outer wall is especially thick.

Fundamental System. The tissues of this system form the main body of the plant, and in the fern have two widely differ-

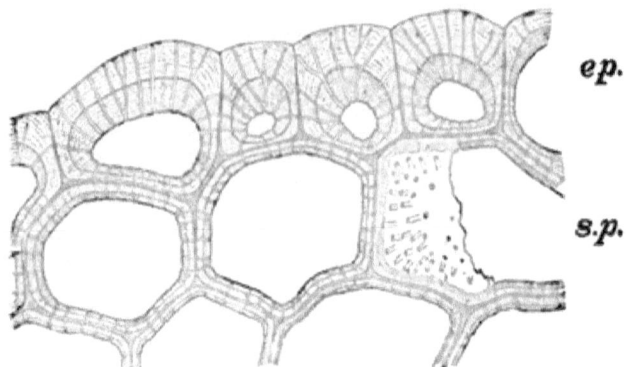

FIG. 49.—Section showing the epidermis (*ep*) and the underlying sclerotic parenchyma (*s.p*) of the rhizome of *Pteris aquilina*. Canals, sometimes branching, are everywhere seen. These served to keep the once-living cells in material connection.

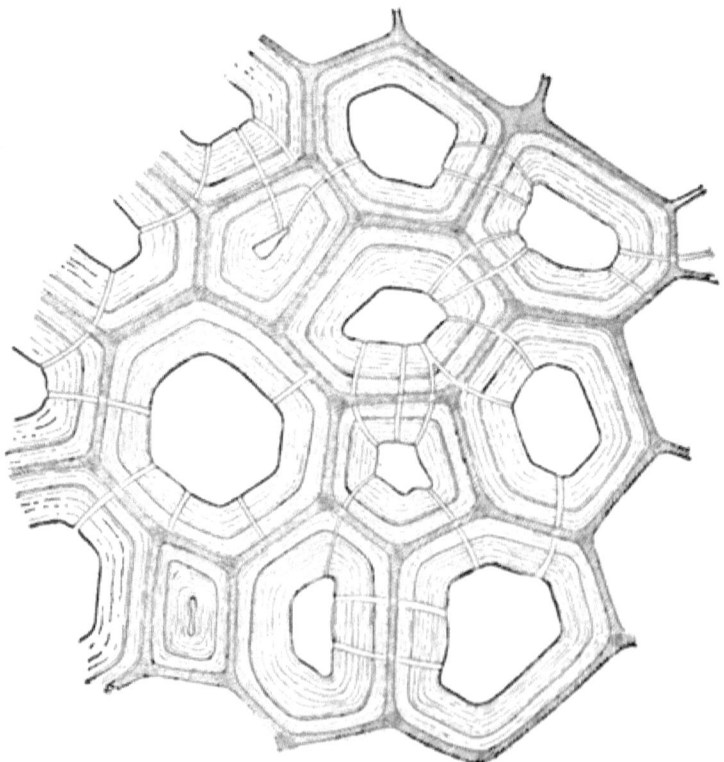

FIG. 50.—Cross-section of sclerotic prosenchyma of the rhizome of *Pteris aquilina*. The enormously thickened walls consist of three layers, are perforated by canals, and are *lignified* or turned into wood.

ent functions. The *fundamental parenchyma* is a kind of storehouse in which matter and energy are stored—mainly in the form of starch, $C_6H_{10}O_5$,—and in which active chemical changes take place. The cells are thin-walled and soft, and are rather loosely joined together, leaving numerous intercellular spaces (Figs. 52, 53). They contain protoplasm and a nucleus, and very numerous rounded grains of starch. This starch is stored up by the plant during the summer as a reserve supply of food —just as hibernating animals store up fat in their bodies for use during the winter. Accordingly, starch increases in quantity during the summer and decreases in the spring when the plant resumes its growth, before the leaves are unfolded. The parenchyma probably has also the function of conducting various substances (especially dissolved sugar) through the plant by diffusion from cell to cell.

The *sclerotic parenchyma* and *sclerotic prosenchyma* (Figs. 49, 50) are dead, and hence play a passive part in the adult vegetal economy. The former co-operates with the epidermis; the latter probably serves in part to support the soft tissues, and to some extent affords a channel for the conveyance of the sap. The sap, however, does not flow through the cavities, but passes slowly along the substance of the porous walls. The cells of both these sclerotic tissues have very thick, hard, brown walls, perforated here and there by narrow canals. The cells of the parenchyma are prismatic or polyhedral; those of the prosenchyma elongated, and pointed at their ends. In both, the protoplasm and nuclei disappear when the cells are fully formed. Towards the apical buds both fade into ordinary fundamental parenchyma.

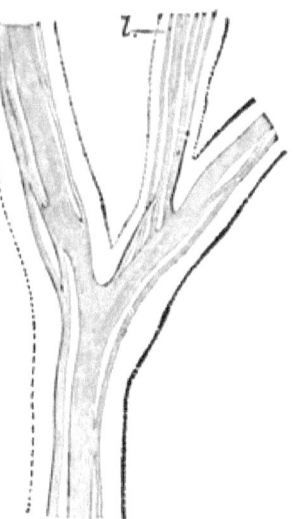

Fig. 51. (After Sachs.)—View of the rhizome, which is supposed to be transparent so as to show the network of the upper fibro-vascular bundles. *l*, a leaf.

Fibro-vascular System. The *fibro-vascular bundles* (p. 115) are long strands or bands of tissue which appear in cross-section as isolated spots (Fig. 48). The bundles are not really isolated, however, but join one another here and there, forming an open network (Fig. 51), which can only be seen in a lateral view of the rhizome. From this network bundles are given off which extend on the one hand into the roots and on the other into the leaves, branching in the latter to form the complicated system of veins to be described hereafter (p. 129).

Each bundle consists of a number of different tissues which, broadly speaking, have the function of conducting sap from one part of the plant to another.

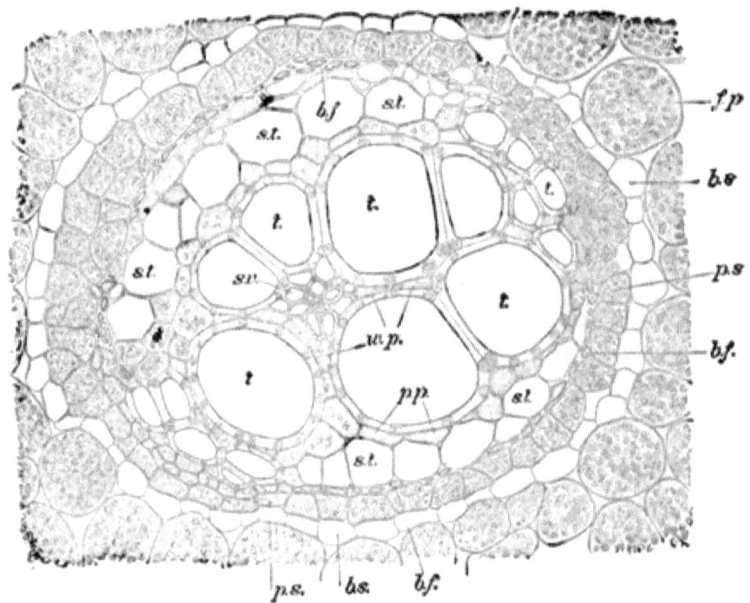

FIG. 52.—Highly magnified cross-section of a fibro-vascular bundle surrounded by the fundamental parenchyma, f.p. t, scalariform tracheids; b.s, bundle-sheath; p.s, phloëm-sheath; b.f, bast-fibres; s.t, sieve-tubes; p.p, phloëm-parenchyma; w.p, wood (xylem) parenchyma; s.v, spiral vessel.

These tissues have the following definite arrangement. Beginning with the outside of a bundle, we find (Figs. 52, 53)—

1. *Bundle-sheath:* a single layer of elongated cells enveloping the bundle, probably derived from and belonging to the fundamental system.

2. *Phloëm-sheath;* a single layer of larger parenchymatous cells containing starch in large quantities.

3. *Bast-fibres;* soft, thick-walled, elongated, pointed cells containing protoplasm and large nuclei.

4. *Sieve-tubes;* larger, soft, thin-walled, elongated cells containing protoplasm and having the walls marked by areas perforated by numerous fine pores (panelled). They join at the ends by oblique panelled partitions (shown in Figs. 52 and 53).

5. *Phloëm-parenchyma;* ordinary parenchymatous cells filled with starch, scattered here and there among the bast-fibres and sieve-tubes.

6. *Tracheids (scalariform)* or "ladder-cells"; occupying most of the central part of the bundle. Their structure calls for some remark. They are empty or air-filled fusiform tubes, whose hard, thick walls are in the young tissue sculptured with great regularity into a series of transverse hollows or pits, which finally become actual holes. The walls of the tracheid are therefore continuous at the angles, but along their plane sur-

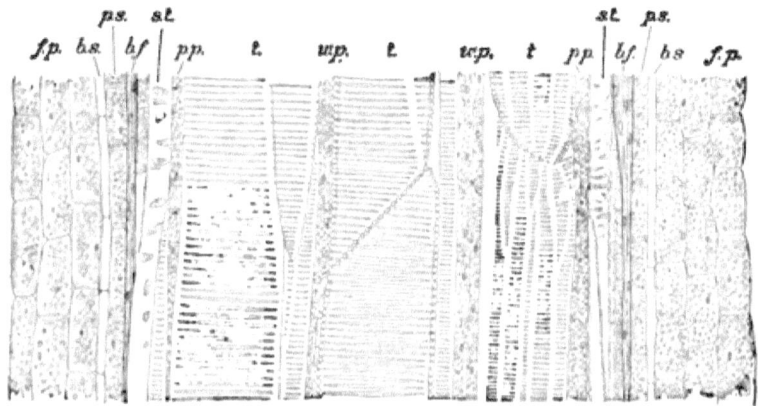

FIG. 53.—Longitudinal section of a fibro-vascular bundle, surrounded by the fundamental parenchyma. *b.f*, bast-fibres; *b.s*, bundle-sheath; *f.p*, fundamental parenchyma; *p.p*, phloëm-parenchyma; *p.s*, phloëm-sheath; *s.t*, sieve-tubes; *t*, scalariform tracheids or ladder-cells; *w.p*, wood-parenchyma.

faces become converted into a series of parallel bars, making a grating of singular beauty. The slits between the bars are not rectangular passages through the wall, but are rather like elongated, flattened funnels, opening outwards. The sides of the funnels are called the *borders* of the *pits;* and pits of this sort are called *bordered scalariform pits* (cf. Fig. 53).

7. *Tracheæ or vessels (spiral)*; scattered here and there among the tracheids, and hardly distinguishable from them in cross-section. They are continuous elongated tubes filled with air, and strengthened by a beautiful close spiral ridge (sometimes double) which runs round the inner face of the wall (Fig. 52).

The tracheids and vessels are of great physiological importance, being probably the main channels for the flow of sap. Sap is water holding various substances in solution. *The water enters by the roots, flows principally through the walls of the vessels and tracheids, and not through their cavities, which are filled with air*, and is thus conducted through the rhizome and upwards into the leaves.

8. *Wood-parenchyma;* cells like those of the phloëm-parenchyma (5) scattered between the vessels and tracheids.

Branches of the Rhizome These repeat in all respects the structure of the main stem. They are equivalent members of the underground part, and differ in no wise, excepting in their origin, from the main stem itself.

Roots. The roots may easily be recognized by their small size and tapering form, and their lack of the lateral ridges of the

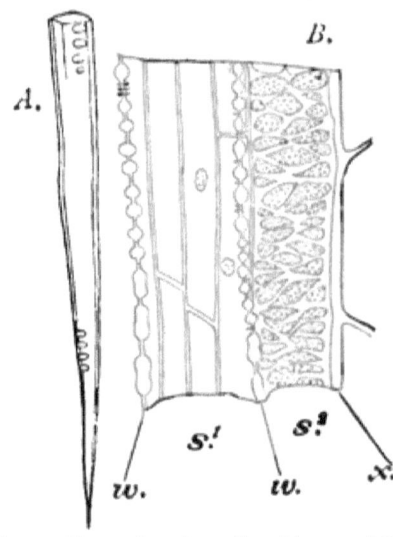

FIG. 54. (After De Bary.)—Sieve-tubes from the rhizome of *Pteris aquilina*, showing: *A*, the end of a member of a sieve-tube; *B*, part of a thin longitudinal section. The section has approximately halved two sieve-tubes, S^1 and S^2, which are so drawn that the uninjured side lies behind. The broad posterior surface of S^2 is seen covered with sieve-plates connecting with another sieve-tube. S^1, on the contrary, abuts by a smooth non-plated surface upon parenchymatous cells which are seen through it. *w*, sections of walls bearing sieve-pits; *x*, section of a non-plated wall abutting upon parenchyma.

stem and branches. They arise *endogenously* from the main stem or its branches, i.e., by an outgrowth of the internal tissues, and not (as in the case of the false roots or *rhizoids* of the prothallium, shortly to be described) by elongation of superficial cells of the epidermis. True roots, of which those of *Pteris* are good examples, arise always as well from the fundamental and fibro-vascular regions, and include all the systems found in the stem itself. Hence cross-sections of *Pteris* roots differ but slightly from those of the stem or the branches, and the root in general is clearly a member of the plant body. As in all true roots, the free end is covered by a special boring tip called the *root-cap*, but this is apt to be lost in removing the specimen from the earth.

The Embryonic Tissue or Meristem of the Rhizome. The mature rhizome remains at the tip nearly undifferentiated into tissues. At this point the epidermis may be distinguished, but it remains very delicate, and the underlying cells continue to grow and multiply, producing continued elongation of the mass. In this way the apical bud is formed. Lateral buds are given off right and left to constitute the embryos of leaves, branches, or roots, which, always retaining their soft and delicate tips, are capable of further growth.

Behind these "growing points" the epidermis and other tissues grow more and more slowly, and soon reach their maximum size, whereupon rapid growth ceases. The power of growth is henceforward mainly confined to the apical buds, and the growing tissue of which they are composed is known as *embryonic tissue* or *meristem*.

The Apical Cell of the Rhizome. Close examination reveals the fact that each apical bud contains a remarkable cell which is especially concerned in the function of growth, viz., the *apical cell*, which lies in a hollow at the apex of the bud. In the apical buds of the rhizome or branches this cell has somewhat the

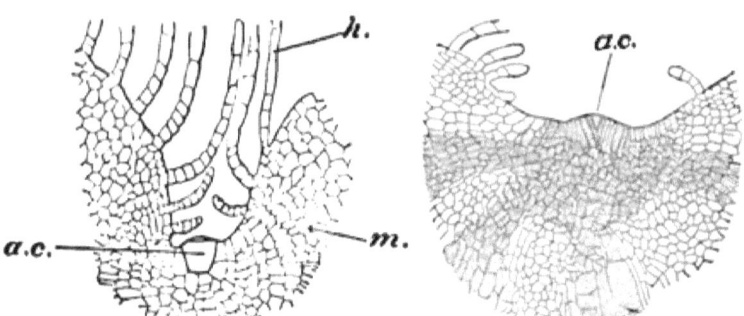

FIG. 55A. (After Hofmeister.)—Apical cell of the rhizome in a vertical longitudinal section. *a.c*, apical cell; *h*, hair; *m*, meristem.

FIG. 55B. (After Hofmeister.)—Apical cell of the rhizome in horizontal longitudinal section. *a.c*, apical cell.

form of a wedge with its base turned forwards and its thin edge backwards, the latter placed at right angles to a plane passing through the lateral ridges. It continually increases in size, but as it grows repeatedly divides so as to cut off cells laterally alternately on its right and left sides. These cells in turn continue to grow and divide, and thus give rise to two similar masses of meristem, which together constitute the apical bud. From the meristem by gradual, though rapid, changes the various tissues of the adult rhizome are differentiated; and longitudinal sections passing through the lateral ridges show the mature tissues fading out in a region of indifferent meristem about the apical cell (Fig. 55B).

The apical cell lies at the bottom of a funnel-shaped depression at the tip of the stem. It is shaped approximately like a thin, two-edged wedge with an arched or curved base turned forwards towards the centre of the funnel-shaped depression. The thin edge of the wedge is directed backwards, and its sides, which are also curved, meet in a vertical plane above and below. A longitudinal section taken through the plane of the lateral

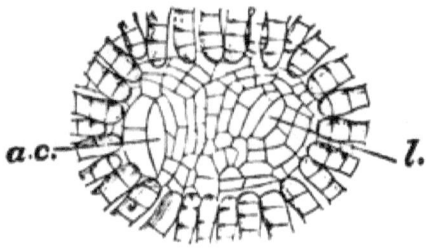

FIG. 56. (After Sachs.)—A vertical transverse section through the *apical cell*, *a.c*, showing a boundary of hairs and a second apical cell, *l*, belonging to a leaf.

ridges therefore shows the apical cell in a triangular form as in Fig. 55B. A section taken at right angles to this—i.e., vertical and longitudinal— shows the cell to be approximately rectangular and quadrilateral (Fig. 55A), while a transverse vertical section shows it in the form of a bi-convex lens (Fig. 56).

The funnel-shaped depression is compressed vertically, and its walls are thickly covered with erect branching hairs, which are closely fastened

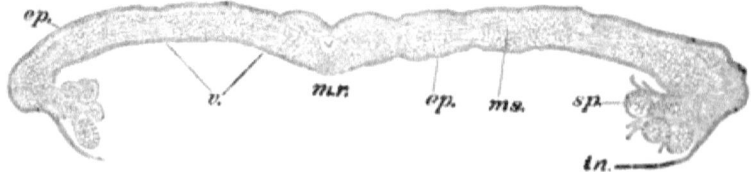

FIG. 57.—Cross-section of an entire fertile leaflet. *m.r*, midrib; *v*, veins; *ep*, epidermis; *ms*, mesophyll; *sp*, sporangia; *in*, indusium.

together by a hardened mucilage secreted by the apical bud. These hairs entirely close the mouth of the funnel and shut off the delicate young portions at its base from the outer world. Protected by these hairs, the end of the stem forces its way through the toughest clay without injury to the delicate bud buried in its apex. (Hofmeister.)

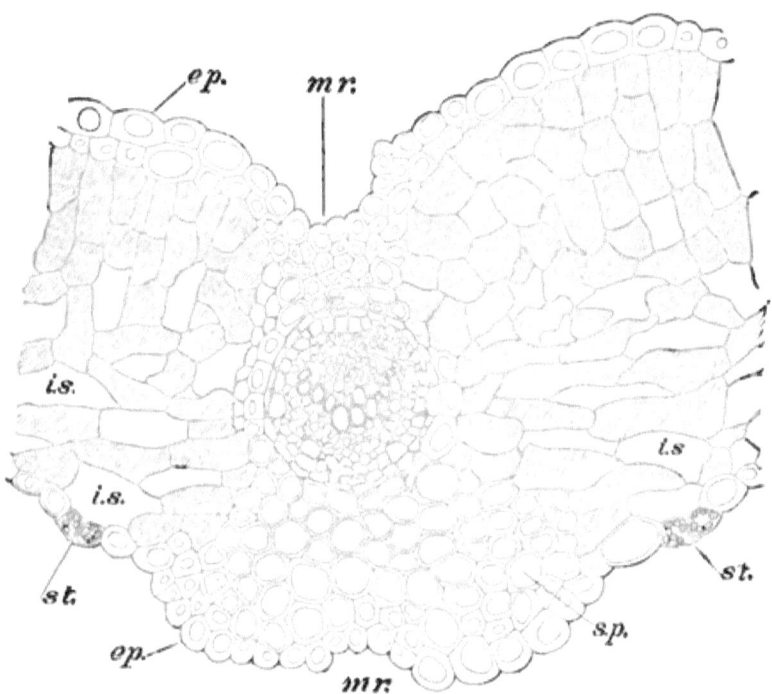

FIG. 58.—Cross-section, still more enlarged, passing through the midrib of a leaflet. In the centre the circular fibro-vascular bundle, supported, especially above and below, by thickened prosenchyma (*p*). On either side the parenchymatous, mesophyll cells (shaded) and the intercellular spaces (*i.s*) opening by stomata (*st*); epidermis (*ep*).

THE AËRIAL PART OF THE BRAKE. THE FROND OR LEAF.

The external form of the leaf has been described on p. 109, and it now remains to consider its internal structure. The lamina is to be regarded as a flattened and altered portion of the stipe, made thin and delicate in order to present a large surface to the light and the air. The stipe, in turn, is a prolongation of the rhizome, so that the whole plant body is a continuous mass, throughout which extend the three systems of tissue virtually unchanged. The transverse and longitudinal sections of the stipe show only minor points of difference from corresponding sections of the rhizome. In the leaf, however, all three systems undergo great changes. The epidermis becomes very thin, delicate, and transparent; the fibro-vascular bundles break up into an extremely fine and complex network forming the

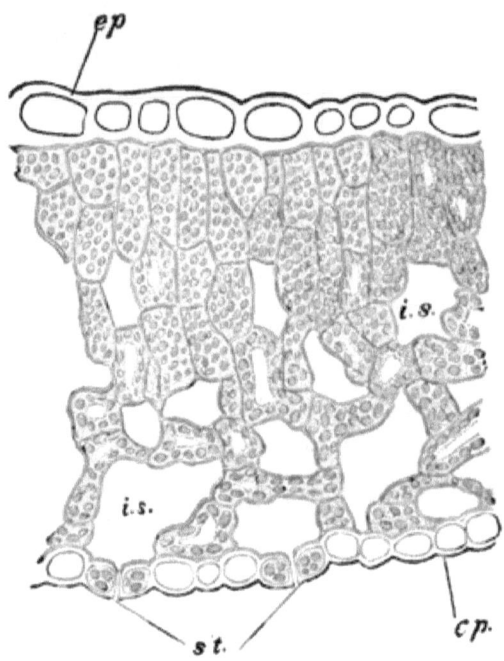

Fig. 59.—Cross-section of part of a leaflet showing the microscopic structure. *ep*, epidermis; *st*, stomata; *i.s.* intercellular spaces between the mesophyll-cells, which are filled with (shaded) chlorophyll-bodies lying in the protoplasm.

veins; the sclerotic tissues become transparent and are found only along the veins. The cells of the fundamental parenchyma alter their form, lose their starch, and become filled with bright-green, rounded bodies, called the *chromatophores* or *chlorophyll-bodies*, which are composed of a protoplasmic basis colored by a pigment known as **chlorophyll**. The green fundamental parenchyma of the leaf is sometimes called the *mesophyll*.

A cross-section of a leaflet (p. 109) is shown in Fig. 57. The finer structure of the leaflet is shown in Figs. 58 and 59. On the outside is the epidermis (*ep*); within, the mesophyll and midrib—the latter composed of thickened epidermal and sclerotic fundamental tissue, and a large fibro-vascular bundle.

The mesophyll, or leaf-parenchyma, consists of irregular cells which are loosely arranged on the lower side, leaving very large intercellular spaces, but are closely packed, and leave few or no intercellular spaces, on the upper (sunny) side. The cells have very thin walls, contain protoplasm and a large central space

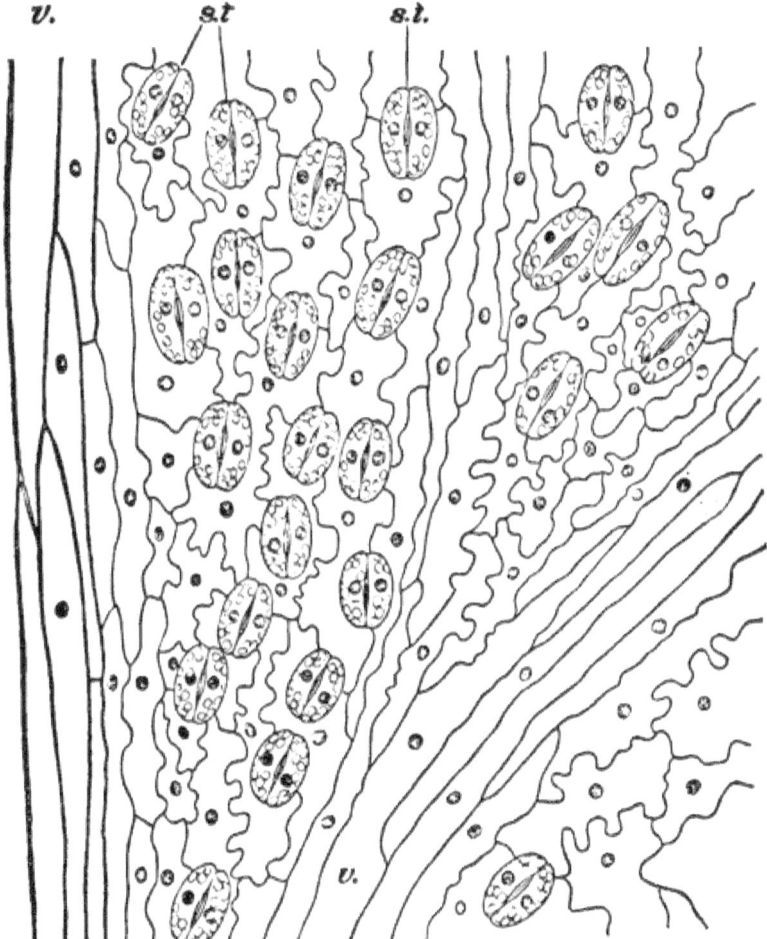

FIG. 60.—Epidermis from the under side of a leaflet, showing wavy cells; elongated (prosenchymatous) cells over the veins; and stomata with their guard-cells. *st*, stomata and guard-cells; *v*, veins covered by thick and prosenchymatous epidermal cells. Intermediate stages between wavy and straight cells are also shown. (Surface view.)

(vacuole) filled with sap, and numerous chlorophyll-bodies imbedded in the protoplasm. These are especially numerous in the upper part of the leaf, as might be expected from their functions in connection with the action of light (see page 147).

The epidermis, or *skin* of the leaf, consists of translucent, greatly flattened cells having peculiar wavy outlines and relatively thick walls (Figs. 58–61). Upon the veins they become elongated, and their walls are considerably thickened, especially upon the midrib (Fig. 58). They generally contain large, distinct nuclei, and often considerable protoplasm. The wavy epidermal cells, particularly in young plants, contain some chlorophyll and starch, though in this respect the fern is somewhat exceptional.

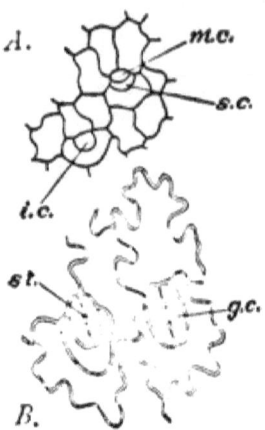

FIG. 61. (After Sachs.)—Epidermal cells of *Pteris flabellata*, showing the development of stomata. *A*, very young epidermal cells; *B*, nearly mature; *i.c.*, initial cell; *m.c.* mother-cell; *s.c*, subsidiary cell; *g.c*, guard-cell; *st*, stoma.

In the rhizome the epidermis forms a continuous layer over the whole surface. In the leaf, however, this is not the case, the epidermis on the lower side being perforated by holes leading into the interior and known as mouths or *stomata* (singular, *stoma*) (Fig. 61). These holes do not pass into the cells, but are gaps or breaks between certain cells of the epidermis, and open directly into the intercellular spaces, of which they are, in fact, the ends. That portion of the intercellular labyrinth which directly underlies the stoma is sometimes called the respiratory cavity. Each stoma is bounded, as in most plants, by two curving *guard-cells*, which are generally nucleated, and, unlike epidermal cells generally, contain abundant chlorophyll-bodies and starch.

The guard-cells are capable of changing their form according to the amount of light, the hygroscopic state of the atmosphere, and other circumstances, and thus open or close the hole or stoma between them. This action is of great importance in the physiology of the plant (transpiration, p. 147).

In *Pteris cretica* and *P. flabellata* the stomata develop as follows: A young epidermal cell is divided by a curved partition into two cells, one of which (Fig. 61) is called the *initial cell* of the stoma (*i.c*). This is again divided by a curved partition into the mother-cell of the stoma (Fig. 61, *m.c*) and a subsidiary cell (Fig. 61, *s.c*).

The mother-cell is then bisected into the two *guard-cells*, and the stoma appears as a chink between them (Fig. 61, *B*).

The *veins* are the fibres or threads which constitute the framework of the leaf. Each consists, essentially, of a small fibro-vascular bundle branching from that of the midrib (Figs. 57, 58, 62). Above and below them the mesophyll and epidermal cells are generally thickened and prosenchymatous, in this way contributing alike to the form and the function of the "vein."

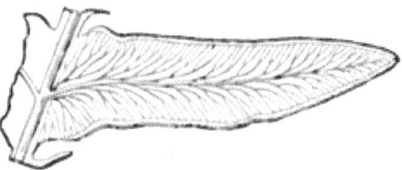

FIG. 62. (After Luerssen.)—Venation of a leaflet of *Pteris aquilina*.

Their arrangement (veining or *venation*) is definite, and depends on the mode of branching of the fibro-vascular strand which constitutes the principal part of the midrib. Secondary strands (nerves) proceed from this at an acute angle, then turn somewhat abruptly towards the edge of the leaflet (or lobe), making an arch which is convex towards the distal extremity of the midrib (Fig. 62).

From this point, after branching once or twice, the delicate veins run parallel to each other to the edge of the leaflet, where they join one another or *anastomose*. This form of venation is known as *Nervatio Neuropteridis*, and is more easily seen in the leaf of *Osmunda regalis* (cf. Luerssen, *Rabenhorst's Kryptogamen-Flora* (1884), III., s. 12).

CHAPTER IX.

THE BIOLOGY OF A PLANT (*Continued*).

Reproduction and Development of the Brake or Fern.

Reproduction. Unlike the earthworm, the fern reproduces both by *gamogenesis* (sexually) and *agamogenesis* (asexually). *Pteris* possesses two modes of asexual reproduction, viz., the detachment of entire branches from the rhizome and the consequent establishment of independent plants, as already mentioned (p. 111), and the formation of "adventitious buds" from the bases of the leaf-stalks (Fig. 46). But besides these the fern has a quite different method of reproduction, in which a process of agamogenesis regularly alternates with gamogenesis (*alternation of generations*). The following brief outline of this important process may help to guide the student through the subsequent detailed descriptions.

Upon some of the leaves are formed organs called *sporangia* (Figs. 57, 63, 64), which produce numerous reproductive cells called *spores*. The spores become detached from the parent and develop into independent plants, the *prothallia* (Fig. 70), which differ entirely in appearance from the fern and ultimately produce male and female germ-cells. The female cell of the prothallium, if fertilized by a male cell, develops into an ordinary "fern," which again produces spores asexually. The formation and development of the spores is evidently a process of *agamogenesis*, and the fern proper is therefore neither male nor female—i.e., it is sexless or *asexual*. The formation and development of the germ-cells, on the contrary, is a process of *gamogenesis;* and the prothallium is a distinct sexual plant, being both male and female (*hermaphrodite* or *bisexual*). In general terms this is expressed by calling the ordinary fern the spore-bearer, or *sporophore*, and the prothallium the egg-bearer, or *oöphore*. The life-history of the fern, broadly speaking, consists therefore in an alternation of the *sporophore* (asexual generation) with the *oöphore* (sexual generation); that is, it consists of an *alternation of generations*. An essentially similar alternation of sporophore with oöphore occurs in all higher plants, though in most cases it is so disguised as to escape ordinary observation.

The Sporangia and Spores. The *sporangia* of *Pteris* (Figs. 63, 64) arise upon a longitudinal thickening of tissue situated on the under side of the leaflets near their edges, and including a marginal anastomosis of the veins. This swelling is known as the *receptacle*. Hairs are not uncommon upon the under side of the leaf, and some are found upon or near the receptacle.

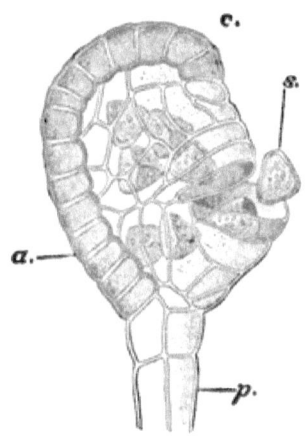

FIG. 63. (After Suminski.)—Sporangium of *Pteris serrulata*. p, pedicel; c, capsule; a, annulus; s, spore.

On the latter arise structures, at first superficially similar to hairs, which become enlarged at the tip, and finally develop into the sporangia. Meanwhile the edge of the leaflet is bent down and under so as to make a longitudinal band of thin tissue composed of epidermis known as the *outer veil* or *indusium* (Fig. 64, *o.i*). A similar thin sheet of epidermis grows down from the under side of the leaf, and passing outwards to meet the former, constitutes the *inner veil* or *true indusium* (Fig. 64, *B, i.i*).

In the V-shaped space thus formed the sporangia are developed.

A superficial (epidermal) cell enlarges and becomes divided into a proximal (basal) cell and a distal (apical) cell (Fig. 65, *a*). The former develops into the future *pedicel* or stalk of the sporangium; the latter gives rise to the head or *capsule* within which the spores are formed (cf. Fig. 63). The pedicel arises from the original pedicel-cell by continued growth and subdivision until it consists of three rows of cells somewhat elongated. The rounded capsule-cell is next transformed by four successive oblique divisions into four plano-convex "parietal cells" and a tetrahedral central cell, the *archesporium*, enclosed by the others. The capsule-cell is thus divided by three planes inclined at about 120° (Fig. 65, *b, c*). A fourth (Fig. 65, *d, e*) passes nearly parallel to the top of the capsule and cuts off from it the central cell or archesporium. In the parietal cells further divisions follow, perpendicular to the surface, while the archesporium gives rise to four intermediate or *tapetal* cells, parallel to the original parietal group (Fig. 65, *g*). The sporangium now consists of a central tetrahedral archesporium bounded by four tapetal cells, which in turn are enclosed by the parietal cells, at this time rapidly multiplying by divisions perpendicular to the exterior. Owing to the peculiar position of the planes of

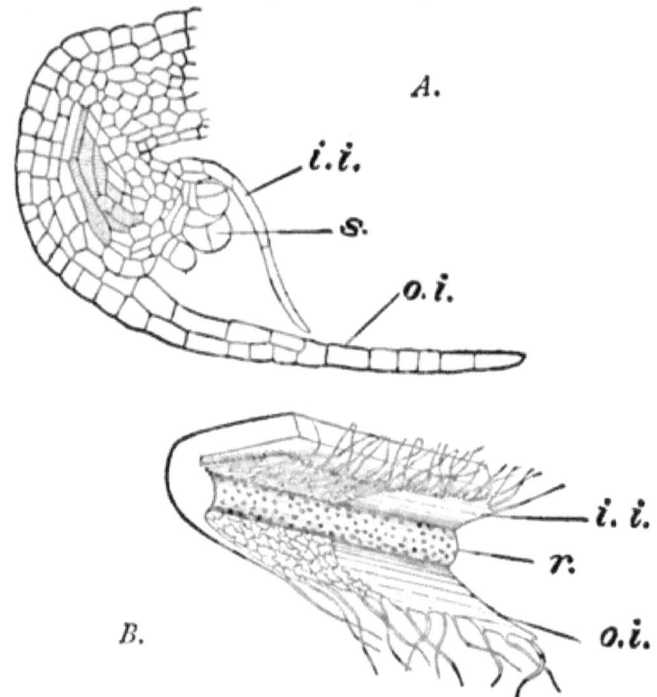

FIG. 64. (From Luerssen, after Burck.)—Indusia and receptacle of *Pteris aquilina;* *B* (diagrammatic), seen from below; *A*, in the section of the edge of a leaflet. *o.i*, outer (false) indusium; *i.i*, inner (true) indusium; *r*, receptacle; *s*, young sporangia.

division the whole capsule is now somewhat flattened, and it becomes still more so by the formation along the edge of a peculiar structure called the *ring* or *annulus*, whose function is the rupturing of the capsule and the liberation of the spores. The annulus is formed by a number of parallel transverse partitions (Fig. 65, *f, h, i, j*), which subdivide the peripheral cells of one edge of the capsule until a certain number of cells have been formed. These then project upon the capsule (Fig. 65, *j*) and form an incomplete ring (Fig. 65, *k*).

Meanwhile the tapetal cells sometimes subdivide so as to form a double row (Fig. 65, *h*), and soon afterwards are absorbed, space being thus left

DEVELOPMENT OF SPORANGIA.

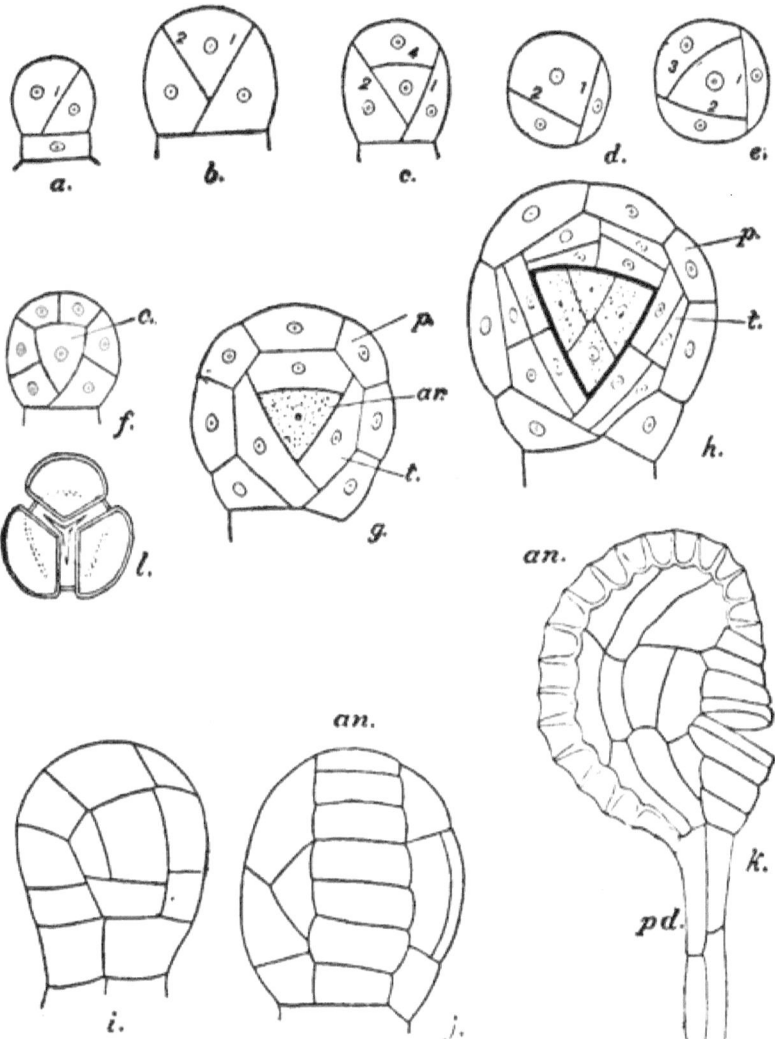

FIG. 65. (After Luerssen.)—Development of the sporangia of *Aspidium Filix mas*, which is closely similar to that of *Pteris*. *a*, the young sporangium standing upon the epidermis-cell from which it has just been divided; *x*, the proximal cell cut off from the sporangium to form the pedicel and support the capsule; *a*, 1, the first partition in the capsule; *b*, 1 and 2, the first and second partitions; *c*, 1, 2, 4, the first, second, and fourth partitions; *d* and *e* are cross-sections of the capsule showing the oblique position of the partitions, and especially that of the third; *f*, a later stage; *g*, the origin of the tapetal cells and the formation of the archesporium; *h*, division of the tapetal cells and the formation of the spore mother-cells; *l*, four spores as they originate in the spore mother-cells; *i, j, k*, the annulus and ripe sporangium, in surface view; *p*, peripheral cells; *ar*, archesporium; *t*, tapetal cells; *an*, annulus.

for the growth and enlargement of the archesporium. The latter now divides—first into 2, then into 4, 8, and finally 16 cells, the *mother-cells of the spores*. These remain for a time closely united, but eventually separate and again subdivide, each into 4 daughter-cells (Fig. 65, *l*). The 64 cells thus formed are the asexual *spores*. In their mature state they have a tetrahedral form and certain external markings, indicated in Figs. 63, 66. Each spore acquires a double membrane, viz., an inner, *endosporium*, delicate and white, and an outer, *exosporium*, yellowish brown, hard, and sculptured over the surface with very close and fine, but irregular, warty excrescences.

Germination of the Spores. Development of the Prothallium. In the brake the spores ripen in July or August and are set free by rupture of the sporangium under the strain exerted by the elastic annulus, as indicated in Fig. 63. Germination of the spores normally occurs only after a considerable period (perhaps not before the following spring); it begins by a rupture of the exospo-

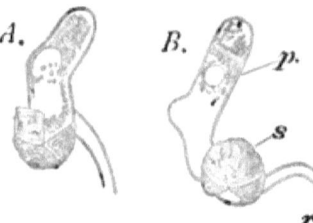

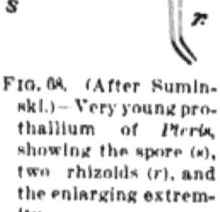

FIG. 66. (After Suminski.) — Single spore of *Pteris serrulata*.

FIG. 67. (After Suminski.)—Germinating spores of *Pteris serrulata*. *A*, in an early stage; *B*, after the appearance of one transverse partition; *s*, spore; *p*, protonema; *r*, rhizoid.

FIG. 68. (After Suminski.)—Very young prothallium of *Pteris*, showing the spore (*s*), two rhizoids (*r*), and the enlarging extremity.

rium which is probably immediately due to an imbibition of water. The spore bursts irregularly along the borders of the pyramidal surfaces, and from the opening thus formed the endosporium protrudes as a papilla filled with protoplasm in which numerous chlorophyll-bodies soon appear.

This papilla is known as the *protonema*, or first portion of the prothallium (Fig. 67). It develops very quickly into a stout cylindrical protrusion divided into cells joined end to end. Close to the spore one or more *rhizoids* are put down from the growing protonema to serve as anchors and roots. At the opposite or distal end longitudinal partitions soon appear (Fig. 68), which speedily convert this portion into a broad flat plate at first only one cell thick, but eventually several cells thick along the median line. This thickening is the so-called "cushion" (see Fig. 70). The whole prothallium is now somewhat spatulate (Fig. 69), but by further growth anteriorly, by an apical cell or otherwise, the wider end becomes still more flattened and heart-shaped or even kidney-shaped. Numerous rhizoids (so-called because they are not morphologically true roots) are put down, and the whole structure assumes approximately the appearance indicated in Fig. 70. The spore-membranes and protonema soon fall away, and the prothallium enters upon an independent existence, being rooted by its rhizoids and having an abundance of chlorophyll. In the broad thin plate of tissue no subdivision into stem and leaf exists, and the plant body closely resembles the "thallus" of one of the lowest plants. Since it is the precursor of the ordinary "fern," it is called the "*prothallus*" or "*prothallium*."

FIG. 69. (After Suminski.) - Older prothallium, showing two rhizoids, three young antheridia, and numerous chlorophyll-bodies.

The cushion forms a prominence on the lower side; upon its posterior part most of the rhizoids are borne.

Sexual Organs of the Prothallium. The prothallia of ferns are as a rule bisexual or hermaphrodite; that is, each individual possesses both male and female organs. But the latter appear somewhat later than the former, and poorly nourished prothallia often bear only male organs, though they will frequently develop female organs also if placed in better circumstances.

The *Antheridia*, or male organs, are hemispherical prominences occurring upon the posterior part and the under side of the prothallium, often among the rhizoids. When fully formed (Figs. 70, 71) an antheridium consists of a mass of rounded cells (*spermatozoid mother-cells*) enveloped by a membrane one cell in thickness.

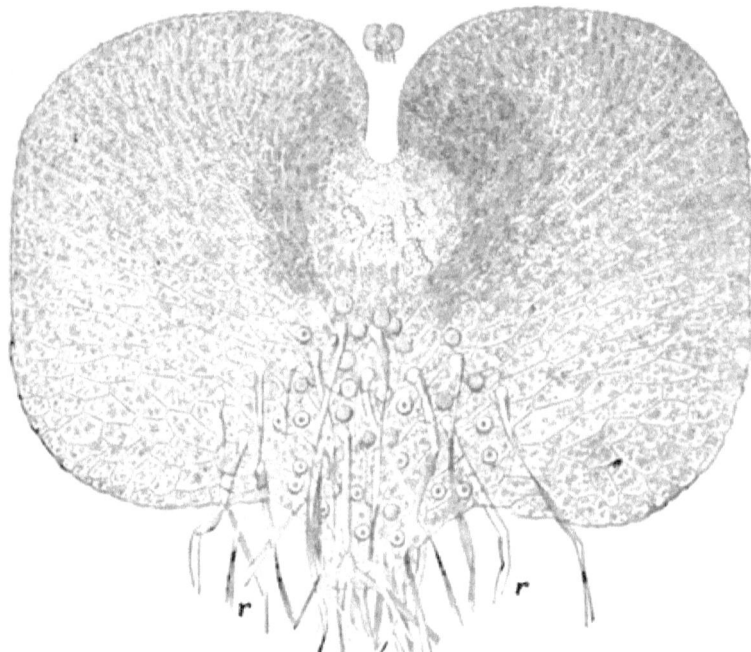

FIG. 70. (After Suminski, slightly modified.) Adult prothallium of *Pteris serrulata* seen from below, showing the rhizoids (*r*) at the posterior end, the depression at the anterior end; the cushion near the latter bearing (in this case) four archegonia. Among the rhizoids are the (spherical) antheridia. The chlorophyll-bodies only are shown in the cells of the broad plate of tissue constituting the prothallium. Just above the anterior depression is seen a prothallium of the natural size.

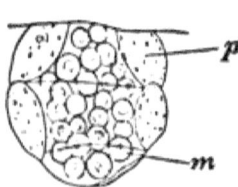

FIG. 71. (After Strasburger.)—Mature antheridium of *Pteris serrulata*. *p*, peripheral cells; *m*, mother-cells of the spermatozoids.

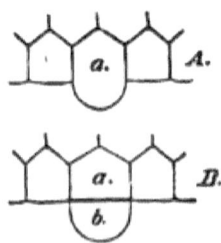

FIG. 72.—Diagram to illustrate the origin of an antheridium. *A*, very young stage; *B*, older; *a*, original epidermal cell enlarged; *b*, mother-cell of the entire antheridium.

MALE GERM-CELLS.

The mode of origin of the mother-cells differs considerably in different ferns, but in all cases is essentially as follows: An ordinary cell on the lower side of the prothallium swells and forms a hemispherical or dome-shaped projection, which is soon separated by a partition from the original cell (Fig. 72). Further divisions then follow in the dome-shaped cell such that a central cell is left, surrounded by a layer of peripheral cells (Fig. 73). By repeated divisions the central cell splits up into the spermatozoid *mother-cells* (Fig. 71).

Within each mother-cell the protoplasm arranges itself in a peculiar spiral body, the *spermatozoid*, which is the *male germ-cell*.

When the mature antheridium is moistened, the peripheral cells swell and thus press out the mother-cells and spermatozoids (Fig. 74). The latter escape from the mother-cells and swim about very actively in the water. They appear as naked single cells, of a peculiar corkscrew shape, and bear upon the finer spirals numerous extremely active cilia (p. 31), by which they are driven swiftly through the water.

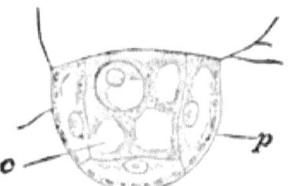

FIG. 73. (After Hofmeister.)—Later stage in the development of an antheridium of *Pteris serrulata*. *p*, peripheral cell; *c*, central cell from which the spermatozoid mother-cells arise.

The *Archegonia*, or female

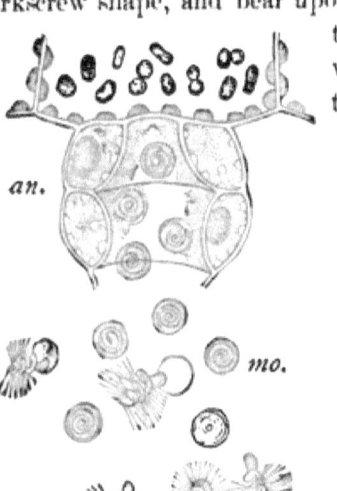

FIG. 74. (After Luerssen.)—Bursting of the antheridium and escape of the spermatozoids. *an*, antheridium; *m.c*, spermatozoid mother-cells; *sp*, spermatozoids.

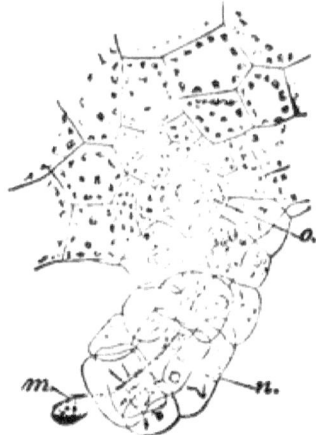

FIG. 75. (After Strasburger.)—Mature archegonium, showing the oösphere (*o*), the neck (*n*), and mucus (*m*) issuing from the mouth of the canal.

organs (Figs. 70, 75), described for the first time by Suminski in 1864, likewise arise from single superficial cells of the prothallium. They are situated almost exclusively upon the cushion near its anterior or apical extremity, and hence at the bottom of the anterior depression (Fig. 70). Since they appear later than the antheridia, they are not likely to be fertilized by spermatozoids descended from the same spore. This phenomenon of maturation of one set of sexual organs of a bisexual individual before the ripening of the other set is a common feature among plants, and is known as *dichogamy*. There is reason to believe that important advantages are gained by thus securing cross-fertilization and preventing self-fertilization or "breeding in and in."

In the development of the *archegonium* the original cell enlarges, becomes somewhat dome-shaped, and divides by transverse partitions into three cells: a proximal, imbedded in the tissue of the prothallium, a middle, and a distal dome-shaped cell (Fig. 76). The fate of the proximal cell is unimportant. The distal cell gives rise by division to a chimney-like structure, *the neck* (Figs. 75, 77), which

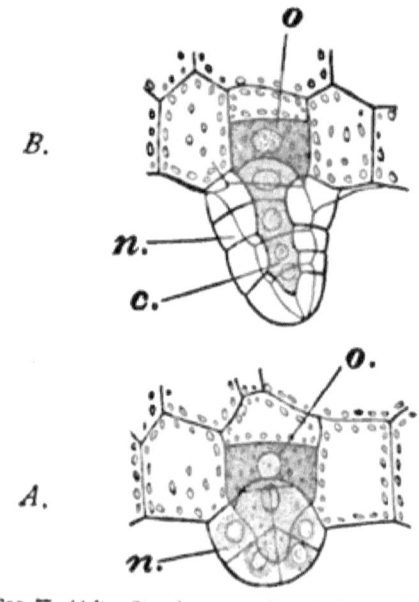

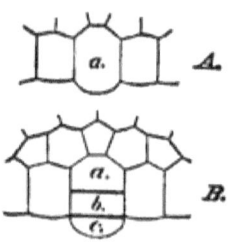

FIG. 76.—Diagram to illustrate the origin of an archegonium. *A*, an early stage; *B*, a later stage; *A, a*, the original epidermal cell enlarged; *B, a*, the basal cell; *b*, the central or canal cell; *c*, the neck-cell.

FIG. 77. (After Strasburger.)—Developing archegonia of *Pteris serrulata*. *A*, young stage; *B*, older; *n*, neck; *c*, canal; *o*, oösphere.

encloses a row of cells (*canal-cells*) derived from the original middle cell (Figs. 75, 77). These afterwards become transformed into a mucilaginous substance filling a canal leading through the neck from the outside to the oösphere (Fig. 77), which also arises from the original "middle" cell at its proximal end. The oösphere is the all-important *female germ-cell* to which the "neck-" and "canal-cells" are merely accessory.

Fertilization or Impregnation. Fertilization, or the sexual act, is performed as follows: Spermatozoids in vast numbers are attracted to the mouths of the archegonia and there become entangled in the mucilage (Fig. 78). In favorable cases one or more work their way down the mucilaginous canal, and at length one penetrates and fuses with the oösphere.

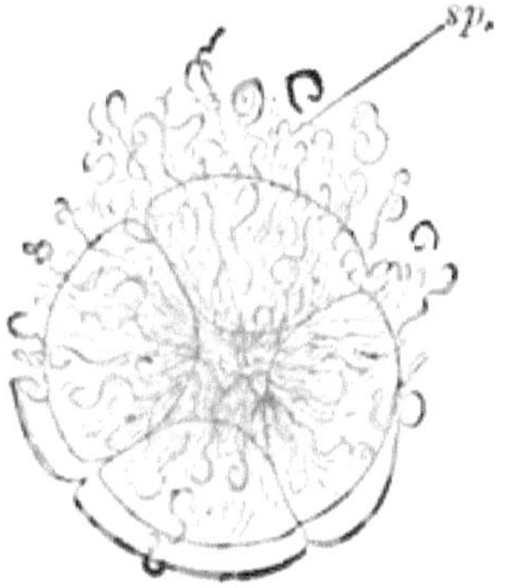

It is known that one spermatozoid is enough to fertilize the oösphere, and probably one only penetrates it; but several are often seen in the mucilaginous canal. It has been shown that the mucilage contains a small amount (about 0.3.)

FIG. 78. (After Strasburger.)— Mouth of an archegonium of *Pteris serrulata*, crowded with spermatozoids striving to effect an entrance.

of malic acid, which probably acts both as an attraction to the spermatozoids and as a stimulus to their movements. Pfeffer has proved that capillary tubes containing a trace of a malate in solution are as attractive to the spermatozoids as is the mucilage in the central canal, and phenomena of this kind (*chemiotaxis*) have recently been shown to be common and highly important.

The entrance of the spermatozoid into the ovum and its fusion with it mark an important epoch in the life-history of the fern. The oösphere is from this instant a new and very different thing, viz., an *embryo*, and is known as the *oöspore*. It is now the first stage of the asexual generation, though it is still maintained for some time at the expense of the sexual generation or *oöphore* (p. 130).

Growth of the Embryo. The oöspore, or one-celled embryonic sporophore (p. 130), now rapidly becomes multicellular by dividing first into hemispheres, then into quadrants, etc. (Fig. 80; compare Fig. 14). The first plane of division is approximately a prolongation of the long axis of the archegonium (Fig. 80). The second is nearly at right angles to it, so that the quadrants may be described as anterior and posterior to the first plane. The fate of the quadrant-cells is of special importance. The lower anterior quadrant as it undergoes further division grows out into the *first root;* the upper anterior quadrant in like manner gives rise to the *rhizome* and the *first leaf.* The mass of cells derived from the two posterior quadrants remains connected with the prothallium as an organ for the absorption of nutriment from the latter, and is inappropriately called the *foot.*

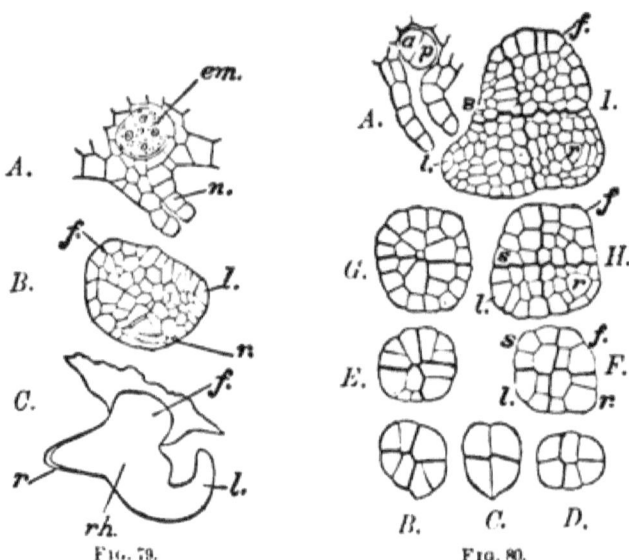

Fig. 79. Fig. 80.

FIG. 79. (After Hofmeister.) - Development of the embryo. *A*, section showing the closed neck (n) and the planes of quadrant division of the oöspore or embryo (em). The fore end of the prothallium is to the right. *B* and *C*, stages of the embryo later than *A*, showing the beginnings of apical growth; *f*, foot; *l*, leaf; *r*, root; *rh*, rhizome.

FIG. 80. (From Luerssen, after Klenitz-Gerloff.) Development of the embryo of *Pteris serrulata.* The figures are optical sections taken vertically in the antero-posterior axis of the prothallium, passing through the long axis of the neck of the archegonium; except *C* and *D*, which are taken at right angles to the others. *A*, *a*, and *p* are the anterior and posterior segments of the oöspore after this has divided into hemispheres. The former (a) forms the stem, the latter (p) the root. *F* shows in a late stage the division of the quadrants, *r* going to form the root, *s* the stem or rhizome, *l* the leaf, and *f* the foot; *r*, *l*, and *s* soon take on apical growth as indicated in *H* and *I*.

In *Pteris serrulata* the development is slightly different. The lower anterior cell becomes the first leaf; the upper anterior becomes the first portion of the rhizome, the lower posterior becomes the primary root, and the upper posterior remains as the "*foot.*"

The several parts now enter upon rapid growth accompanied by continued cell-multiplication, until a stage is reached repre-

GROWTH AND DIFFERENTIATION. 141

sented in *C*, Fig. 79. A stage somewhat later than this, with its attachment to the prothallium, is shown in Fig. 81. After this the leaf grows upwards into the air, the root downwards into the earth, and the young fern begins to shift for itself. Eventually it reaches a condition shown in Figs. 82 and 83. The prothallium remains connected with the young fern for some time, and may readily be found in this condition attached to flower-pots in hot-houses, etc. But sooner or later it falls off, and the young fern enters upon an entirely independent existence. The appearance of the plant and the shape of the leaf do not always at first resemble those of the adult fern; growth is also more rapid at first, several leaves (7–12) being developed successively in the first year (p. 112).

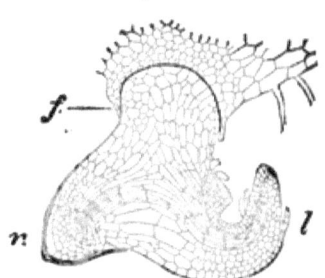

FIG. 81. (After Hofmeister.)—Young embryo of *Pteris aquilina*, showing its attachment to the prothallium by the foot; *l*, leaf; *f*, foot; *r*, first root.

Differentiation of the Tissues. In the earliest stages the tissue is nearly or quite homogeneous, i.e., meristemic. But very early in development, as the leaf turns upwards and the root

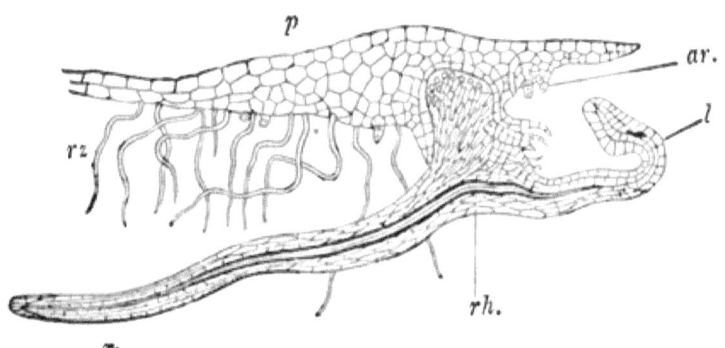

FIG. 82. (After Sachs.)—Older embryo of maidenhair-fern (*Adiantum*) attached to the prothallium. Seen in section. *l*, leaf; *r*, first root; *rh*, beginning of the rhizome; *p*, prothallium; *rz*, rhizoids; *ar*, archegonia.

downwards, changes take place, which lead directly to a differentiation into the three great systems of tissue—epidermal, fibrovascular, and fundamental. The epidermal and fundamental systems take on almost at once the peculiarities which have already been noted in the adult, p. 117. The fibro-vascular system of tissues is differentiated a little later. Different as the tissues of the three systems are, it is plain from their mode of origin that all are fundamentally of the same nature because of their descent from the same ancestral cell; hence every cell in the plant partakes more or less completely of the nature of every other cell. The resemblances are primary and fundamental, the differences secondary and derived. And what is true of the fern in this respect is equally true of all other many-celled organisms.

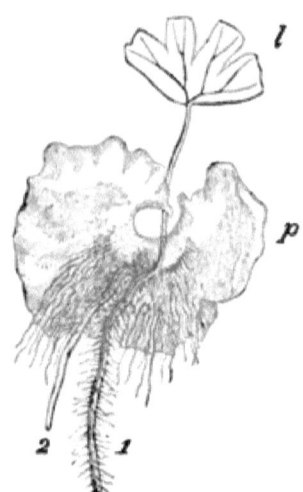

Fig. 83. (After Sachs.)—Young maidenhair-fern (Adiantum) attached to the prothallium, p. l, leaf: 1, 2, the first and second roots.

Course of the Fibro-vascular Bundles. Certain features of the disposition and course of the fibro-vascular bundles in the embryo and in the adult may conveniently be studied at this point. From the point of junction of the bundles of the first leaf and first root (Figs. 79, 81, 82) is developed one central bundle traversing the young rhizome and sending branches into the new leaves and roots until 7–9 leaves have been formed. After this time the rhizome forks, and the course of the fibro-vascular bundles in each fork is henceforwards compound. A lateral depression appears in the central bundle of each stem, rapidly increases in depth, and soon divides the bundle into two, one upper and one lower, which are best recognized in old specimens (Fig. 48). When the forked shoots have reached a length of about three inches, these bundles send out at a small angle towards the periphery thinner, forked branches which soon unite again to form a network near the epidermis. The uppermost of these branches, which passes in the median line above the axile bundles, is usually somewhat more fully developed, and almost as broad as the latter. This structure is generally retained in the mature rhizome (Fig. 48, x). The number of peripheral bundles may be as great as twelve in the cross-section. They anastomose in the vicinity of the place of insertion of each frond, and thus form a hollow, cylindrical network, having elongated meshes; but no connecting branches between them and the two axile bundles are found anywhere in the rhizome. The latter follow an entirely isolated course within the creeping stem;* branches from them

---

* See, however, De Bary, *Comp. Anat. Phanerogams and Ferns*, p. 295. Oxford, 1884.

enter the leaves, and it is only inside the leaf-stalk that these ramifications are met by branches from the peripheral network. The bundles of the roots arise only from the peripheral bundles, but those of leaves, as already said, receive branches from both axillary and peripheral bundles. Two thick brown plates (*sclerotic prosenchyma*) lie between the inner and outer systems of bundles, and are only separated from one another at the sides by a narrow band of parenchyma. They are often joined on one side or even on both, in the latter case forming a tube which separates the two systems of bundles. (Hofmeister.)

Apogamy. Apospory. In rare cases, e.g., in *Pteris cretica*, the ordinary alternation of generations in the life-cycle of ferns is abbreviated by the omission of the sexual process, and the immediate vegetative outgrowth of the sporophore from the prothallium (*apogamy*). In other cases there is an omission of the spore stage, and immediate vegetative development of the oöphore from the frond (*apospory*). (cf. Farlow, *Quart. Journ. Mic. Science*, 1874 ; De Bary, *Botan. Zeitung*, 1878; Druery, etc., *Journ. Royal Mic. Soc.*, 1885, pp. 99 and 491.)

CHAPTER X.

THE BIOLOGY OF A PLANT (Continued).

The Physiology of the Fern.

The brake, like the earthworm, is a limited portion of organized matter occupying a definite position in space and time. It is bounded on all sides by material particles, some of which may be living, but most of which are lifeless. The aerial portion is immersed in and pressed upon by an invisible fluid, the atmosphere, while the underground portion is sunk in a denser medium, the earth, which likewise acts upon it. At the same time the fern reacts upon the air and the earth, maintaining during its life an equilibrium which is disturbed and finally gives way as the life of the plant draws to a close.

The Fern and its Environment. Those portions of space, earth, and air which are nearest to the brake constitute its immediate environment. But in a wider and truer sense the environment includes the whole universe outside the plant. To perceive the truth of this it is only necessary to observe how profoundly and directly the plant is affected by rays of light which travel to it from the sun over a distance of many millions of miles, or how extremely sensitive it is to the alternations of day and night or of summer and winter. The plant is fitted to make certain exchanges with its environment, drawing from it certain forms of matter and energy, and returning to it matter and energy in other forms. Its whole life is an unconscious struggle to wrest from the environment the means of subsistence; death and decay mark its final and unconditional surrender.

Adaptation of the Organism to its Environment. We can distinguish in *Pteris* as clearly as in *Lumbricus* the adaptation of the organism to its environment. The aerial part of *Pteris* must be fitted to make exchanges with, and maintain its life in, the atmosphere, while the underground part must be similarly "adapted" to the soil in which it lives.

The aerial part displays admirable adaptation in its stalk, which rises to a point of vantage for procuring air and light, and in its broadly spreading top, which is covered by a skin, tough and impervious, to prevent undue evaporation and consequent desiccation, yet translucent, to allow the sun's rays to reach the starch-making tissue within. The rhizome also, with its pointed terminal buds, its elongated roots, armed with boring tips, and its thick, fleshy parenchyma for the storage of food, is admirably adapted to its own special surroundings. In order to realize this, we have only to imagine the fern to be inverted, the aerial portion being planted in the earth, and the underground portion lifted into the air and exposed to the winds and sunshine. Under these circumstances the want of adaptation of the parts to their respective environments would speedily become apparent.

Yet different as these parts now are, they have originally sprung from the same cell. More recently they were barely distinguishable in a mass of tissue, part of which turned upwards into the air, while another part turned downwards into the earth. But as development went on, the aerial and underground parts were progressively differentiated, thus becoming more and more perfectly adapted to the peculiar conditions by which each is surrounded.

Thus it appears that the harmony between every part of the plant and its environment is brought about, as in the animal, by a *gradual process* in the history of each individual. We can here clearly see also the functional adaptation of the plant to changing external conditions. The environment of *Pteris* changes periodically with the regular alternation of summer and winter, and the plant also undergoes a corresponding periodic change of structure in order to maintain its adaptation to the environment. During the summer the aerial part is fully developed, and, as a result of its activity, starch is accumulated in the rhizome. At the approach of winter the aerial part dies, and the plant is reduced to the underground part safely buried in the soil. During the winter and spring the starch is gradually consumed, and the aerial part is put forth again as the aerial environment becomes once more favorable to it. The plant, therefore, like the animal, possesses a certain *plasticity* which enables it to adapt itself to gradually changing conditions of the environment.

A little consideration will show that every function or action of living things may be regarded as contributing to the same great end, viz., harmony with the environment; and from this point of view life itself has been defined as "*the continuous adjustment of internal relations to external relations.*" *

Nutrition. The fern does work. In pushing its stem through the soil, in lifting its leaves into the air, in moving food-matters from point to point, in building new tissue, in the process of reproduction, and in all other forms of vital action, the plant expends energy. Here, as in the animal, the immediate source of energy is the living protoplasm, which, as it lives, breaks down into simpler compounds. Hence the need of an income to supply the power of doing work.

The Income. The income of the fern, like that of the earthworm, is of two kinds, viz., matter and energy, but *unlike that of the worm it is not chiefly an income of foods, but only of the raw materials of food.* Matter enters the plant in the liquid or gaseous form by *diffusion*, both from the soil through the roots (liquids), and from the atmosphere through the leaves (gases). We have here the direct absorption into the body proper of food-stuffs precisely as the earthworm takes in water and oxygen. Energy enters the plant, to a small extent, as the potential energy of food-stuffs, but comes in principally as the kinetic energy of sunlight absorbed in the leaves. The table on p. 147 shows the precise nature and the more important sources of the income.

Of the substances, the solids (salts, etc.) must be dissolved in water before they can be taken in. Water and dissolved salts continually pass by diffusion from the soil into the roots, where together they constitute the sap. The sap travels throughout the whole plant, the main though not the only cause of movement being the constant *transpiration* (evaporation) of watery vapor from the leaves, especially through the stomata. The gaseous matters (carbon dioxide, oxygen, nitrogen) enter the plant mainly by diffusion from the atmosphere, are dissolved by the sap in the leaves and elsewhere, and thus may pass to every portion of the plant.

The Manufacture of Foods—especially Starch. *Pteris* owes its power of absorbing the energy of sunlight to the *chlorophyll-*

---

* Spencer, *Principles of Biology*, vol. i. p. 80. N. Y., Appleton, 1881.

bodies or *chromatophores*; for plants which, like fungi, etc., are devoid of chlorophyll are unable thus to acquire energy. Entering the chlorophyll-bodies, the kinetic energy of sunlight is applied to the decomposition of carbon dioxide (CO_2) and water (H_2O). After passing through manifold but imperfectly known processes, the elements of these substances finally reappear as starch ($C_6H_{10}O_5$) often in the form of granules imbedded in the chlorophyll-bodies, and free oxygen, most of which is returned

INCOME OF *PTERIS*.

Matter.	Whence Derived.
Carbon.	Mainly from the atmosphere as carbon dioxide (CO_2), but perhaps partly from dissolved organic matters (food).
Hydrogen.	Mainly from the soil as water (H_2O), but perhaps partly from organic foods.
Oxygen.	Mainly from the soil as water (H_2O) and from the air as free oxygen.
Nitrogen.	Mainly from the soil * as nitrates or ammonium compounds, or organic foods.
Sulphur.	Mainly from the soil as sulphates.
Other elements.	Mainly from the soil as various salts.
Energy.	
Kinetic.	Mainly from the sunlight through the leaves.
Potential.	Perhaps to a limited extent in food materials *via* the roots.

to the atmosphere. Thus the leaf of *Pteris* in the light is continually absorbing carbon dioxide and giving forth free oxygen.

Carbon dioxide and water contain no potential energy, since the affinities of their constituent elements are completely satisfied. Starch, however, contains potential energy, since the molecule is relatively unstable, i.e., capable of decomposition into simpler, stabler molecules in which stronger affinities are

---

* It has been generally believed that plants are unable to make use of free atmospheric nitrogen, but recent investigations have disproved this view for certain species.

satisfied. And this is due to the fact that in the manufacture of starch in the chlorophyll-bodies the kinetic energy of sunlight a was expended in lifting the atoms into position of vantage, thus endowing them with energy of position. In this way some of the radiant and kinetic energy of the sun comes to be *stored up* as potential energy in the starch. In short, *Pteris*, like all green plants, is able by co-operation with sunlight to use simple raw materials (carbon dioxide, water, oxygen, etc.) poor in energy or devoid of it, and out of them to *manufacture food*, i.e., complex compounds rich in available potential energy. We shall see hereafter that this power is possessed by green plants alone; all other organisms being dependent for energy upon the potential energy of ready-made food. This must in the first instance be provided for them by green plants; and hence without chlorophyll-bearing plants animals (and colorless plants as well) apparently could not long exist.

The plant absorbs also a small amount of kinetic energy, independently of the sunlight, in the form of heat; this, however, is probably not a source of vital energy, but only contributes to the maintenance of the body temperature.

Circulation of Foods. It is chiefly in the green (chlorophyll-bearing) parts of the plants, and in the presence of sunlight, that food-manufacture goes on. Somehow, then, the water absorbed by the roots must be transported to the leaves, and the starch made in the leaves must be conveyed to the subterranean tissues. Exactly how these transfers of material are effected is uncertain, but there is reason to believe that they take place mainly by the slow processes of diffusion. It is certain that no distinct organs of circulation or distribution, such as the blood-vessels of the earthworm, exist in the fern.

Metabolism. Starch, as has just been seen, is first formed in the chlorophyll-bodies. But the formation of starch, all-important as it is, *is after all only the manufacture of food* as a preliminary to the real processes of nutrition. These processes must take place everywhere in ordinary protoplasm; for it is here that oxidations occur and the need for a renewal of matter and energy consequently arises (cf. pp. 32 and 33). Sooner or later the starch grains are changed into a kind of sugar (*glucose*, $C_6H_{12}O_6$), which, unlike starch, dissolves in the sap, and may thus be easily transported to all parts of the plant. Wherever there is need for new protoplasm, whether to repair previous waste or to supply materials for growth, after absorption into the cells the elements of the starch (or glucose) are, by the living protoplasm, in some unknown way combined with nitrogen and sulphur (probably also with salts, water, etc.), to form proteid matter. The particles of this newly-formed compound are incorporated into the protoplasm (by "intus-susception," p. 4) and, in some way at present shrouded in mystery, are endowed with the properties of life. We do not know how long they may remain in the living state, but sooner or later they are oxidized, and, as a result of the oxidation, that energy is set free which enables the fern to do work and prolong its existence. The oxidized products are afterwards eliminated (excreted) from the cells.

If a larger quantity of starch is formed in the chlorophyll bodies than is immediately needed by the protoplasm for purposes of repair or growth, it may be re-converted into starch after journeying as glucose through the plant, and be laid down as "reserve starch" in the parenchyma of the rhizome, or elsewhere. Apparently, when this reserve supply is finally needed at any point in the plant, it is again changed to glucose and transported thither. It is probable that new leaves and new tissues generally, are always formed in part from this reserve starch, and not solely from newly-formed starch.

In dealing with the metabolism of the fern we may safely assume, as we have done already for the earthworm, a constructive phase (*anabolism*) and a destructive phase (*katabolism*); but these terms represent merely probable events, not known facts.

The Outgo. The outgo, like the income, is of two kinds, matter and energy, but it cannot be so readily tabulated.

The plant suffers annually a great loss both of matter and of potential energy in the production of spores and in the autumnal dying-down of the fronds. But matter also leaves the plant daily as carbon dioxide (in small quantities), water, and oxygen, both by diffusion through the epidermis and by transpiration through the stomata. Strictly speaking, the term outgo should be restricted to the output of matter which has at some time actually formed a part of the living protoplasm; hence it does not apply to the oxygen, which is simply given off in the manufacture of starch, or to the bulk of the water of evaporation, which passes straight through the plant without undergoing any chemical change. Energy likewise leaves the plant continuously both as *heat* and in the *doing of mechanical work*, both of which are involved in every vital act.

Respiration. It has been remarked that in the light (i.e., when manufacturing starch) *Pteris* takes in carbon dioxide and gives off free oxygen. But if the plant be deprived of light, as at night, the reverse is true, and the plant takes in a small amount of oxygen and gives off a corresponding amount of carbon dioxide. This latter process is the true *breathing* or *respi-*

PTERIS AQUILINA.
(Balance-Sheet of Nutrition.)

INCOME.	OUTGO.
Matter.	*Matter.*
Foods,	Carbon dioxide,
Inorganic salts,	Water,
Carbon dioxide,	Excreted substances,
Water,	Reproductive germs,
Free oxygen.	Leaves, etc.,
	Free oxygen — from decomposition of carbon dioxide in light.
Energy.	*Energy.*
Sunlight absorbed by chlorophyll,	Work performed.
Potential energy in foods.	Heat.
	Potential energy in cast-off matters, reproductive germs, etc.

Balance in favor of the living *Pteris:*
Matter.
   Tissues, protoplasm, starch, cellulose, chlorophyll, etc.
Energy.
   Potential energy in organic matters.

ration of the plant, and it must not be confounded with that taking in of carbon dioxide and giving off of oxygen which is an incident in the manufacture of starch. Respiration goes on in the light also, probably with greater energy than in darkness, but it is then largely obscured by the other and more conspicuous process. We have seen that energy is set free in living matter by a decomposition of its own substance, which is really a process of oxidation or combustion, where free oxygen plays an important part (p. 32, Chap. III.); hence the absorption of free oxygen in respiration. Among the products of the combustion, water and carbon dioxide are the most important; and this is the origin of the carbon dioxide given off. It will appear beyond that precisely the same action takes place in the respiration of animals, and that all living things breathe or respire in essentially the same way.

It was for a long time believed that a leading difference between plants and animals lay in the fact that the former give off oxygen and absorb carbon dioxide, while the latter give off carbon dioxide and absorb oxygen. But it is now known that both give off carbon dioxide and both require oxygen, and that only the chlorophyll-bearing parts of green plants are endowed with the special function of decomposing carbon dioxide and water and manufacturing starch—as a result of which they do (but in the light only) give off oxygen as a kind of incidental- or by-product.

INTERACTION OF THE FERN AND ITS ENVIRONMENT.

The actions of the environment upon the fern have already been sufficiently dwelt upon (p. 144). It still remains, however, to consider the actions of the fern upon the environment. These are partly physical, but mainly chemical. By pushing its fronds into the air and slowly thrusting its rhizome, roots, and branches through the soil, the atmosphere and the earth are alike displaced. But it is by its chemical activity that it most profoundly affects its environment. Absorbing from the latter water, salts, carbon dioxide, and other simple substances, as well as sunlight, it produces with them a remarkable metamorphosis. It manufactures from them as raw materials *organic matter* in the shape of starch, fats, and even proteids. These it gives back to the environment in some measure during life, and surrenders wholly after sudden death. But the most striking fact is that the fern is on the whole constructive and capable of producing and accumulating compounds rich in energy. In this respect it is unlike the earthworm (p. 104) and is typical of green plants in general. Thus, while animals are destroyers of energized compounds, green plants are producers of them. Animals, therefore, in the long run are absolutely dependent on plants; and animals and colorless plants alike upon green plants. But it must never be forgotten that most plants are enabled to manufacture organic from inorganic matter by virtue of the chlorophyll which they contain. Without this they are powerless in this respect. (See, however, p. 197).

Physiology of the Tissue-Systems. The *epidermal* tissues serve as the sole medium of exchange between the inner parts of the plant and the environment; they are also protective, and in certain regions are useful for support. The function of reproduction also falls upon these tissues, as is shown by the development of the sporangia, antheridia, and archegonia.

The *fibro-vascular* tissues serve in part as a supporting skeleton, for which function their richness in prosenchyma and their firm continuity admirably adapt them. An equally important function, however, is their *conductivity*, since they serve for the transportation of the water for evaporation by the leaf (*transpiration*), and for the movement (through the sieve-tubes) of the undissolved and indiffusible proteids. The *fundamental tissues* are devoted either to sharing the special duties of the other systems, as in the case of the sclerotic parenchyma abutting upon the epidermal tissue in the rhizome (p. 119), and the sclerotic prosenchyma which appears to behave like the fibro-vascular tissues; or to nutritive and metabolic functions, as in the mesophyll (p. 126) and the parenchyma of the rhizome.

The Physiology of Reproduction. It is not known whether the brake ever dies of old age. Barring accidents, growth at the apical buds seems to be unlimited, keeping pace with death of the hinder parts of the rhizome (p. 111). But whether the individual dies or not, ample provision against the death of the race is made in the act of reproduction. Although reproduction appears to be useless to the individual, and even entails upon it serious annual losses of matter and energy, yet to this function every part of the plant directly or indirectly contributes. The reproductive germs are carefully prepared; are provided with a stock of food sufficient for the earliest stages of development; and are endowed with the peculiar powers and limitations of *Pteris aquilina*, which influence their life-history at every step and are by them transmitted in turn to their descendants. They are living portions of the parent detached for reproductive purposes; they contain a share of protoplasm directly descended from the original protoplasm of the spore from which the parent came; and thus they serve to effect that "continuity of the germ-plasm" to which we have already referred in dealing with the earthworm. In short, reproduction is the supreme function of the plant. If we may paraphrase the words of Michael Foster, the oösphere is the goal of individual existence, and life is a cycle, beginning with the oösphere and continually coming round to it again.

Comparison of the Fern and the Earthworm. To the superficial observer the fern and earthworm seem to have little or nothing in common, except that both are what we call alive. But whoever has studied the preceding pages must have perceived beneath manifold differences of detail a fundamental likeness between the plant and animal, not only in the substantial identity of the living matter in the two but also in the construction of their bodies and in the processes by which they come into existence. Each arises from a single cell which is the result of the union of two differently-constituted cells, male and female. In both the primary cell multiplies and forms a mass of cells, at first nearly similar but afterwards differentiated in various directions to enable them to perform different functions, i.e., to effect a physiological division of labor. In both, the tissues thus provided are associated more or less closely into distinct organs and systems, among which the various operations of the body are distributed. And in both the ultimate goal of individual existence is the production of germ-cells which form the starting-point of new and similar cycles.

This fundamental likeness extends also to most of the actions (physiology) of the two organisms. Both possess the power of adapting themselves to the environments in which they live. Both take in various forms of matter and energy from the environment, build them up into their own living substance, and finally break down this substance more or less completely into simpler compounds by processes of internal combustion, setting free by this action the energy which maintains their vital activity. And, sooner or later, both give back to the environment the matter and energy which they have taken from it. In other words, both effect an exchange of matter and of energy with the environment.

Nevertheless the plant and the animal differ. They differ widely in form, and the plant is fixed and relatively rigid, while the animal is flexible and mobile. The body of the plant is relatively solid; that of the animal contains numerous cavities.

The plant absorbs matter directly through the external surface; the animal partly through the external and partly through an internal (alimentary) surface. The plant is able to absorb simple chemical compounds from the air and earth, and kinetic energy from sunlight; the animal absorbs, for the most part, complex chemical compounds and makes no nutritive use of the sun's kinetic energy. By the aid of this energy the plant manufactures starch from simple compounds, carbon dioxide, and water; the animal lacks this power. The plant can build up proteids from the nitrogenous and other compounds of its food; the animal absolutely requires proteids in its food. And by manufacturing proteids within its living substance, the plant is relieved of the necessity of carrying on a process of digestion in order to render them diffusible for entrance into the body.

Still, great as these differences appear to be at first sight, all of them, with a single exception, fade away upon closer examination. This exception is the *power of making foods*. Plants and animals differ in form because their mode of life differs; but a wider study of biology reveals the existence of innumerable animals (corals, sponges, hydroids, etc.) which have a close superficial resemblance to plants, and of many plants which resemble animals, not only in form, but also in possessing the power of active locomotion. The stomach of the worm, as shown by its development, is really a part of the general outer surface which is folded into the body; and the animal, like the plant, therefore, really absorbs its income over its whole surface —oxygen through the general outer surface, other food-matters through the infolded alimentary surface.

In like manner it is easy to show that not one of the differences between the plant and animal is fundamentally important save the *power of making foods*. The worm must have complex ready-made food including proteid matter. So must the fern; but the fern is able to *manufacture* this complex food out of very simple compounds. In terms of energy, the worm requires ready-made food rich in potential energy; the fern, aided by the sun's energy, can manufacture food from matters devoid of energy.

Hence it appears, broadly speaking, that the fern by the aid of solar energy is constructive, and stores up energy; the earthworm is destructive, and dissipates energy. And this difference becomes of immense importance in view of the fact that the fern is typical in this respect of all green plants, as the earthworm is typical of all animals.

It will hereafter appear that even this difference, great as it is, is partly bridged over by colorless plants like yeast, moulds, bacteria, etc., which have no chlorophyll, are therefore unable to use the energy of light, and hence must have energized food. But these organisms do not, like animals, require proteid food, being able to extract all needful energy from the simpler fats, carbohydrates, and even from certain salts. When we consider that the distinctive peculiarities of animals can thus be reduced to the sole characteristic of dependence on proteid food, we cannot doubt that the differences between plants and animals are of immeasurably less importance than their fundamental likeness.

———

It has been the object of the foregoing chapters to give the student a general conception of organisms, whether vegetal or animal; of their structure, growth, and mode of action; of their position in the world of matter and energy, and of their relations to lifeless things. With this preliminary knowledge as a basis, the student is prepared to take up the progressive study of other organisms, selected as convenient types or examples. It is convenient to begin with low and simple forms of life and work gradually upwards; and it is especially desirable to do so because there is reason to believe that this course corresponds broadly with the path of actual evolution.

CHAPTER XI.

THE UNICELLULAR ORGANISMS.

It has been shown in the foregoing pages that the complex body of an adult fern or earthworm, or of any of the higher forms of life, originates from a single cell of microscopic size. This cell—the fertilized ovum or oösphere—gives rise by division to new cells which in their turn divide, generation after generation, until a full-grown *body* is formed, composed of myriads of cells. But the process of cell-division does not in this case go as far as complete cell-*separation*, and the cells do not acquire a complete individuality. They do, it is true, acquire a certain independence of structure and function; and their individual characteristics may even depart widely from those of neighboring cells (differentiation). Nevertheless they remain closely united by either material or physiological bonds to form one body. The body is not, however, to be regarded as merely an assemblage of independent individual cells. *The body is the individual;* its more or less perfect division into cells is only a basis for the physiological division of labor; of which cell-differentiation is the outward expression.

All this is true, however, only in the higher types. At the bottom of the scale of life there is a vast multitude of forms in which the body consists, not of many cells but of only one, and is therefore comparable in structure not to the adult fern or earthworm, but to the germ-cells from which these arise. Such forms are known as *unicellular* organisms, in contradistinction to the *multicellular*. Like other cells the unicellular organisms multiply by division, but division is followed sooner or later by complete separation; the daughter-cells become entirely distinct and independent individuals, and do not remain permanently associated. In them a true multicellular body, therefore, is never formed; *the cell is the individual, and the body is unicellular.*

Nevertheless the one-celled organism performs all of the characteristic operations of life. A single mass of protoplasm, a single cell, unites in itself the performance of all the various elementary functions which in the multicellular forms are distributed among many cells, differentiated into divers tissues and organs. The unicellular forms are therefore in a physiological sense as truly "organisms" as the multicellular forms; and in many cases the unicellular body shows a very considerable degree of differentiation among its parts. But the unicellular forms are organisms reduced to their lowest terms: they present us with the problems of life in their most rudimentary form. Hence they may afford a kind of key to the more elaborate organization of the higher types.

We shall find among unicellular forms representatives both of animals and of plants, and to a detailed examination of some of these we may now proceed.

CHAPTER XII.

UNICELLULAR ANIMALS (*Protozoa*).

A. Amœba.
(The Proteus Animalcule.)

General Account. *Amœba* is a minute organism occasionally found in stagnant water, in the sediment at the bottom of ponds and ditches, on the surface of water-plants, in damp earth, in organic infusions of various kinds—almost anywhere, in short, in the presence of moisture, organic matter, and other favorable conditions. There are many species of *Amœba*, some living in salt water, others in fresh. One of the largest and commonest fresh-water forms is *Amœba Proteus*, which forms the subject of this account.*

Amœba occurs in an active or *motile* state, and a quiescent or *encysted* state. When active the body consists (Fig. 84) of a minute naked mass of protoplasm which in the case of *large specimens* is barely visible to the naked eye—i.e., half a millimetre ($\frac{1}{50}$ inch) or less in length. This mass creeps, or rather flows, actively about by the continual protrusion of lobes or processes of its own substance, known as *pseudopodia*. These may be put forth from any part of the surface and again merged into the general mass; the body therefore continually changes its shape, and hence the name " Proteus."

When the body is well extended the protoplasm is seen to consist of a clear peripheral substance, the *ectoplasm*, and a central substance, the *entoplasm*, filled with coarse granules which give the body a highly characteristic granular appearance sometimes described as a " gray color." Within the ectoplasm the more fluid entoplasm freely flows, as if confined in a tube or

---

* Other common forms are the smaller *A. radiosa* and *A. verrucosa*. The large *A.* (*Pelomyxa*) *villosa* and *A* (*Dinamœba*) *mirabilis* are not infrequent. See Leidy, Fresh-water Rhizopods of North America.

THE PROTEUS ANIMALCULE. 159

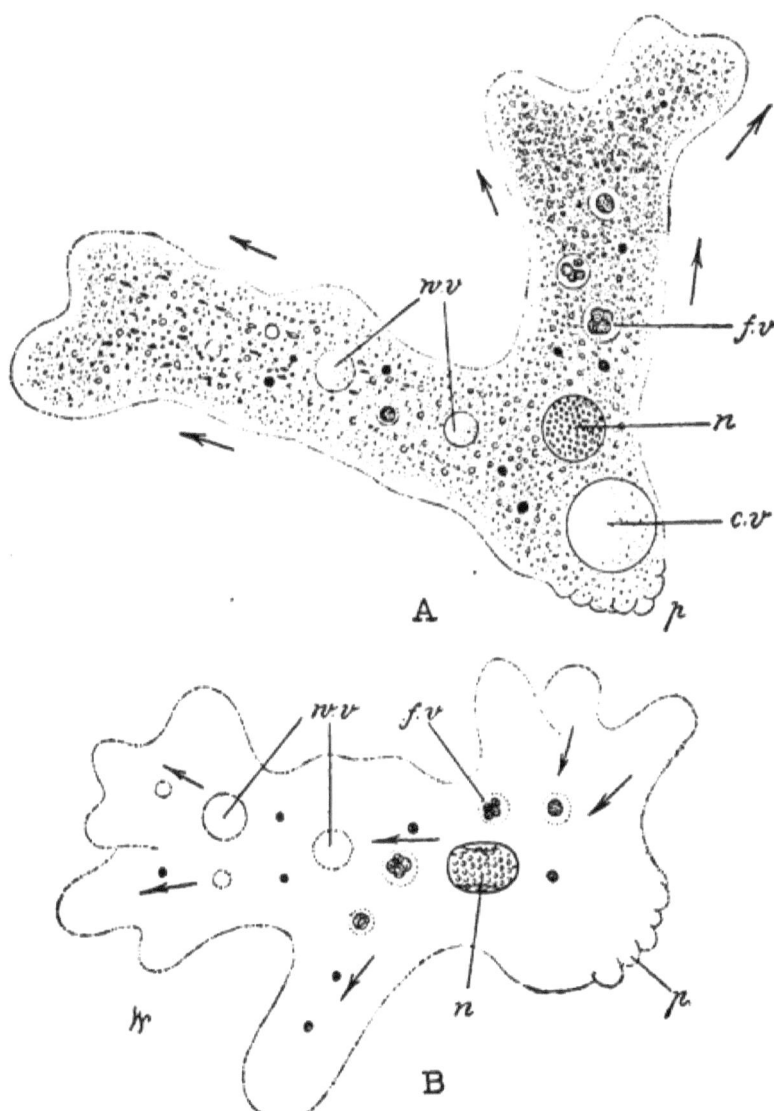

FIG. 84.—*Amœba Proteus*, from life × 300. The arrows indicate the direction of the protoplasmic currents; *n*, nucleus; *c.v*, contractile vacuole; *f.v*, food-vacuole; *w.v*, water-vacuole. *A* shows the texture of the protoplasm. *B* is an outline of the same individual four minutes later; the upward currents at the right of Fig. *A* have stopped, reversed, and the main flow is now towards the left.

sac, but the two substances are not separated by any definite boundary-line, and pass imperceptibly into one another. The external boundary of the body is formed by the outermost limit of the ectoplasm. There is no membrane, and the body is quite naked. Nevertheless the protoplasmic mass shows no tendency to mix with the surrounding water, and perfectly maintains its integrity; it is an individual.

The formation of a pseudopod begins by the bulging out of the ectoplasm to form a rounded prominence at some point on the surface. Into its interior a sudden gush of entoplasm then takes place and a steady outward stream ensues, the entoplasm pushing the ectoplasm before it, and the substance of the body flowing into the pseudopod. The whole substance of the body may thus flow onward into the pseudopod, which meanwhile forms new pseudopods, and so the entire animal advances in the direction of the flow; or, the pseudopod after attaining a certain size may be withdrawn into the body by reverse (centripetal) currents, the main body having meanwhile flowed onward in another direction.

As a rule, the new pseudopodia are put forth near one end of the body (hence called "anterior"), and the general direction of advance is therefore fairly constant, not vague and indefinite, as is often stated. The direction of flow fluctuates, however, about a certain mean, being continually diverted this way or that by the formation of new pseudopodia. Those which do not form directly in the line of march either merge little by little with the advancing ones, or are withdrawn by reversed currents into the body. In the latter case they often leave shrivelled wart-like remnants, and a group of similar warts is usually found near the "posterior" end of the body (Fig. 84, p). Definite changes in the general direction of advance are effected by the diversion of the main current into lateral pseudopodia.

Amœba feeds upon minute plants and animals or other organic particles. There is no mouth, and food-matters are bodily ingulfed (at no definite point) by the protoplasm which closes up beyond them.* The indigestible remains are passed out in

---

* This mode of cellular alimentation is of frequent occurrence in some cells of multicellular, as well as in unicellular, animals. Cells exhibiting it are known as *phagocytes* (eating-cells), and the process is referred to as *phagocytosis*. It is obviously only a prelude to intra-cellular digestion.

an equally primitive fashion, usually at some point near the "posterior" end. Besides solid food-stuffs *Amœba* takes in a certain quantity of water (along with minute quantities of inorganic salts dissolved in it), and it also breathes, by taking in (mainly by diffusion) the free oxygen dissolved in the water and giving off carbon dioxide.

Such is *Amœba* in its active phase. The *quiescent* or *encysted* state is entered upon under conditions not thoroughly understood, but probably of an unfavorable nature, such as the

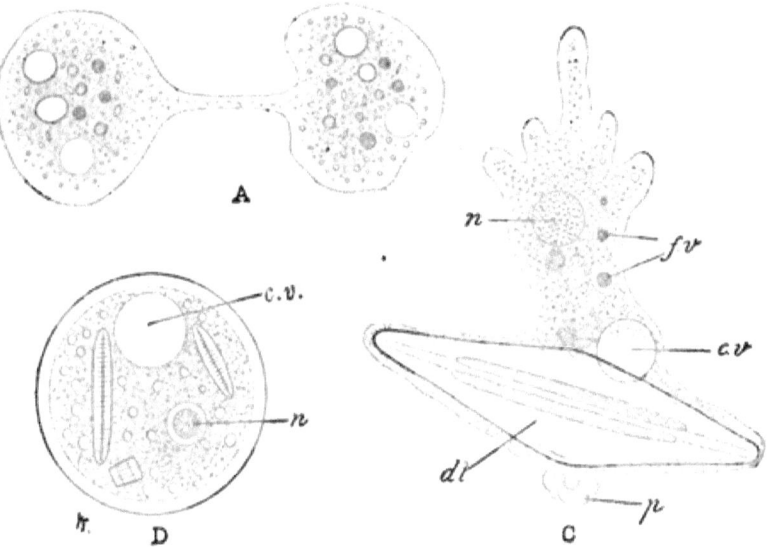

FIG. 85.—*A*, *Amœba* dividing by fission, nucleus not seen (after Leidy). *C*, *Amœba* after a full meal consisting of a large diatom (*dt*). (After Leidy). Letters as in Fig. 84. *D*, Encysted *Amœba*, containing food-matters (after Howes).

lack of food, drying up of ponds, and the like. The pseudopodia are withdrawn, movement ceases, the body becomes spherical and surrounds itself with a tough membrane (cell-wall) (Fig. 85, *D*). The animal takes no food and all of its activities are nearly suspended. It is like an animal asleep or hibernating, and in this state it may long remain. Protected by its membrane it is able to resist desiccation, and upon the evaporation of the surrounding water it may, as a particle of "dust," be transported by the winds, even to a great distance. When again placed under favorable conditions the protoplasm bursts its envelope, crawls forth from it, and reassumes its active phase.

Structure. Lying in the entoplasm, usually near the posterior extremity, is a nucleus (n, Fig. 84), having the form of a bi-concave disk and largely made up of coarse granules of *chromatin* (cf. p. 23). *Amœba* is therefore at once a single cell and a unicellular organism, morphologically equivalent to a single tissue-cell of a higher animal or to the germ-cell from which every multicellular form arises. The *body* of *Amœba* is a one-celled body.

The *protoplasm* (cytoplasm) consists of a clear basis, and (in the case of the entoplasm) of innumerable granules extremely diverse in form and size, and frequently differing in character in different individuals. Often they are in the form of rhomboidal crystalline bodies; in other cases they are rounded or irregular. Their precise chemical composition is uncertain, but they are probably complex organic compounds, a product of metabolism and serving as reserve food-matter.*

Vacuoles. The protoplasm often contains rounded vacuoles of which the three following kinds may be distinguished:

(*a*) *Water-vacuoles* ($w.v.$, Figs. 84, 85), filled with water, lying in the entoplasm and carried along in its currents.

(*b*) *Food-vacuoles* ($f.v.$), also lying in the entoplasm, containing the solid food-matters that have been ingulfed. Within them digestion takes place. When this process is completed they approach the exterior—usually at some point near the posterior end—the outer wall breaks through, and the innutritious remnants are cast out, the ectoplasm closing up the breach immediately afterwards. Thus *Amœba* has no mouth, alimentary canal, or anus, but the general mass of protoplasm plays the *rôle* of all three.

(*c*) *Contractile vacuole* ($c.v.$). Usually single, sometimes double, lying near the posterior end, and filled with liquid. This is sharply distinguished from the other vacuoles by its rhythmical pulsation, expanding (*diastole*) and contracting (*systole*) at regular intervals. During the diastole the vacuole slowly fills with liquid which drains into it from the surrounding protoplasm. At the *systole*, which is very sudden, this liquid is forcibly expelled to the exterior through an opening that breaks

---

* In some species of *Amœba* the entoplasm may also contain innumerable grains of sand taken in from the exterior, but this is not the case in *A. Proteus*.

through the ectoplasm, and immediately afterwards disappears. The contractile vacuole is almost certainly to be regarded as a simple kind of excretory apparatus, the water which collects in it containing in solution various products of destructive metabolism which are thus passed out of the body.*

Reproduction. However abundant the food-supply *Amœba* never grows beyond a certain maximum limit. After this limit has been attained the animal sooner or later divides by "*fission*" into two smaller *Amœba* (Fig. 85, *A*). Thus the existence of an individual *Amœba* is normally terminated, not by death, but by resolution into two new individuals. This process is the simplest possible form of agamogenesis, and *Amœba* is not known to multiply in any other way.† The fission of *Amœba* is a process essentially of the same nature as the division of ordinary tissue-cells, a division of the nucleus preceding that of the cytoplasm. Whether the division of the nucleus is of the indirect type (i.e., passes through the phenomena of karyokinesis) is not known by direct observation, but there is some reason to believe that it is so. In any case the successive fissions of *Amœba* are directly comparable with the successive cleavages of the egg of a metazoön (p. 25). The progeny of the *Amœba*, however, separate and form independent individuals, while those of the egg-cell remain intimately associated to form a single multi-cellular individual. Morphologically, therefore, a metazoön is comparable not with a single *Amœba*, but with a multitude of *Amœba*.

Physiology. The possible simplicity of animal structure is well shown in *Amœba*, which is morphologically an animal reduced to its lowest terms. Its physiological operations are correspondingly primitive and rudimentary; and by an analysis of them we may discover what is essential and fundamental in the physiology of animals in general. A survey of the various activities of *Amœba* shows that these may all be reduced to a few *fundamental physiological properties* of the protoplasm,‡ as follows:

---

* It may be recalled that the cavity of the nephridium in the earthworm is intra-cellular, like a vacuole (p. 60).

† It has been asserted that *Amœba* conjugates and also that it multiplies by endogenous division; but the evidence on both these points is inconclusive.

‡ It is hardly necessary to remark that in common with all English-speaking biologists we are indebted to Foster for the first comprehensive elaboration of the "fundamental physiological properties" as exhibited by *Amœba*.

(1) *Contractility*, by means of which motion is effected. This appears most clearly when the animal is stimulated by a sudden jar, or by an electric shock, which causes the body to contract into a ball. This property, precisely like the contraction of a muscle (p. 27), is the result of a molecular rearrangement, accompanied by chemical changes, which causes a change of form in the mass without altering its bulk. The action of the contractile vacuole is due to the contractility of the surrounding protoplasm; and in like manner the currents which cause the protrusion and withdrawal of pseudopods, and so the locomotion of the animal as a whole, are produced by localized contractions of the peripheral layer of protoplasm which drive onwards the more fluid central parts.

(2) *Irritability (including Co-ordination)*, or the power to be affected by, and to respond to, changes or "stimuli" acting upon or within the protoplasm. The change of shape following the application of an electric shock is actually effected by contractility, but the power to be affected by the shock and to arouse contractility, is irritability. To this property the animal owes its power of performing adaptive actions in response to changes in the environment, and also its power to co-ordinate the various actions of its own body. To illustrate: It is a remarkable fact that *Amœba* is able to discriminate between nutritious and innutritious matters, ingulfing the former, but rejecting the latter. Physiologically this discrimination is *a difference of response to different stimuli*—hence a phenomenon of irritability. Again, the various actions (movements, etc.) of *Amœba*, despite their apparently vague character, are *co-ordinated* to form a definite whole; and co-ordination may be regarded as a phenomenon of irritability, changes in one part serving as stimuli to other parts and being brought into orderly relation with them. The property of irritability lies at the base of all nervous activity in higher forms (cf. p. 67) and is concerned in many other actions.

(3) *Metabolism*, the most fundamental of all vital actions, since it lies at the root of all, is the power of waste and repair—the destructive chemical changes in protoplasm (*katabolism*) whereby energy is set free, and the constructive actions (*anabolism*) through which new protoplasm is built and potential energy is stored (cf. p. 33). There is every reason to believe

that the metabolic phenomena of *Amœba* are, broadly speaking, similar to those of higher animals. The katabolic changes are in the long run processes of oxidation, and although their products have not yet been definitely ascertained in *Amœba*, there can be no doubt that they consist mainly of carbon dioxide, water, and some form of nitrogenous matter (urea or a related substance). Most of these waste matters are believed to be passed out (*secretion, excretion*) by means of the contractile vacuole, but probably carbon dioxide leaves the body by diffusion through the general surface (*respiration* in part).

The materials for the constructive process (*anabolism*) are derived from organic food-matters—bodies or fragments of plants and animals taken as food in the process of *alimentation*, and *absorption* from the water and the inorganic salts dissolved in it, and from the free oxygen that enters by diffusion through the general surface (*respiration* in part). Proteid matter is an indispensable constituent of the food, and *Amœba* is therefore an animal.

Alimentation, absorption, secretion, digestion, and circulation, all of which are only the prelude to metabolism, but which in the higher animals are assigned to different organs, tissues, and cells, are here performed by one and the same cell. The capture of solid food here requires its entrance into the cell; and the fact that proteids cannot be absorbed by diffusion necessitates intracellular digestion which in turn necessitates cellular defaecation. It will be observed that while there is no localized or permanent mouth or anus, the whole surface of the cell is potentially mouth or anus. In short, the protoplasm here exhibits not the physiological division of labor, but its absence.

(4) *Growth and Reproduction.* Logically there is in the case of *Amœba* no good ground for a distinction between these processes and metabolism; for reproduction is directly or indirectly an effect of growth, and growth is simply an excess of anabolism over katabolism. Practically, however, the distinction is necessary; for the tendency of living things to run in cycles of growth and reproduction is one of their most obvious and characteristic features.

Here, as in all protoplasmic structures, growth takes place throughout the mass, by *intussusception* (p. 4), not by the additions of superficial layers, as in the case with growth by *accretion* (inorganic bodies, e.g., crystals). Under favorable condition of nutrition this process exceeds the destructive process so that the body increases in size up to a limit, at which fission takes place. What determines this limit is unknown, but the cause is perhaps in some way connected with the geometrical principle that the volume of the cell increases as the cube of its diameter, whereas the surface, by which it absorbs nutriment, and otherwise comes into relation with the outside world, increases only as the square of the diameter. No great increase in size, therefore, is possible without destroying the normal equilibrium of the cell and hence the periodic reduction of size by division. This principle is, however, too general to be of much value. Different species of *Amœba* differ in size-limit, and the immediate cause lies in some subtle relation between organism and environment that cannot at present be made out. It is not known whether or not the *Amœba* ever dies of old age.

These "fundamental physiological properties" of protoplasm lie at the basis of all physiology, and will be found applicable to all forms of life whether vegetal or animal.

Related Forms. *Amœba* is a representative of a very extensive class of Protozoa known as *Rhizopoda*, all characterized by the power to form pseudopodia, and agreeing with *Amœba* in many other respects. One of the commonest fresh-water forms is the genus *Arcella* (Fig. 86, *C*), which even in the active phase is surrounded by a brown horny membrane ("shell") perforated by a large rounded opening through which pseudopodia are protruded. *Difflugia* (Fig. 86, *B*), also a common fresh-water form, builds about itself a beautiful vase-shaped or retort-shaped shell composed of sand-grains, or even, in some cases, of diatom-shells. In *Actinophrys*, or the "sun-animalcule" (Fig. 86, *A*), the pseudopodia are stiff needle-shaped processes radiating in every direction.

Among the marine forms two groups (orders) are of especial interest and importance; viz., the *Foraminifera*, which secrete a calcareous shell perforated by numerous pores, and the *Radiolaria*, which have a siliceous shell. Many of these forms float at the surface of the water, and their cast-off shells have in former times accumulated at the bottom in such enormous quantities as to form beds of chalk in the case of Foraminifera, while the remains of Radiolaria have made important contributions to the formation of siliceous rocks.

FRESH-WATER RHIZOPODS. 167

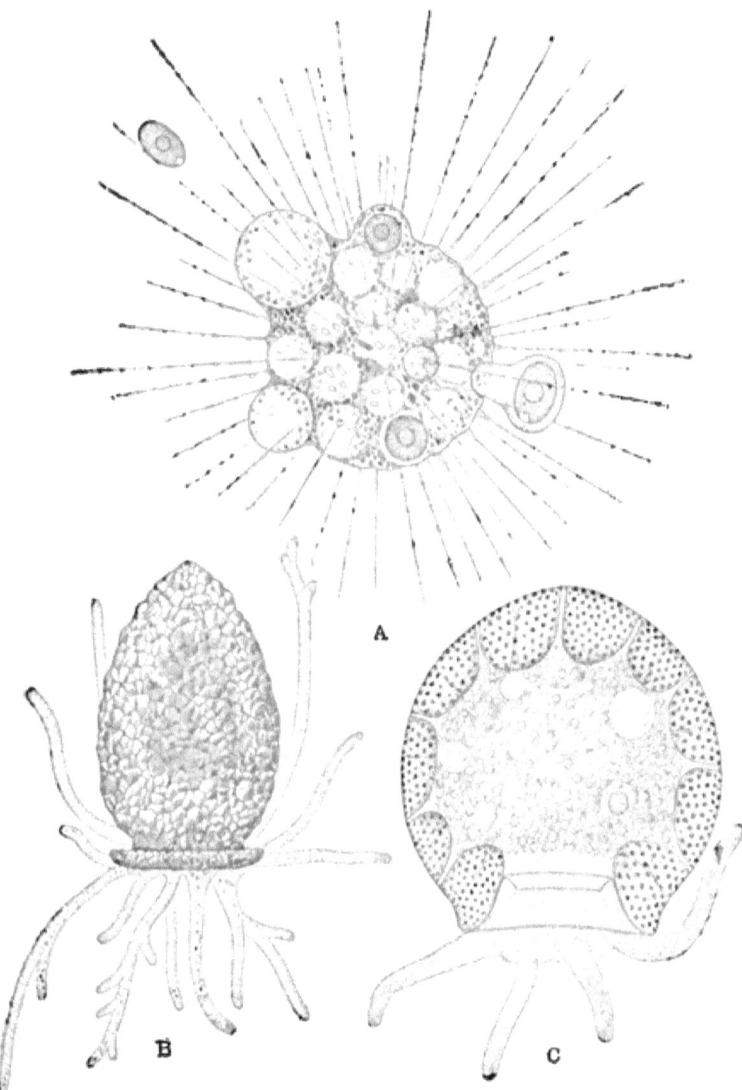

FIG. 86.—Group of common fresh-water Rhizopods (after Leidy). *A, Actinophrys sol*, the "sun-animalcule," filled with vacuoles and containing three food-bodies (zoöspores of an alga); a fourth is just being ingulfed. The nucleus is not seen.
B, Difflugia urceolata, with shell built of sand-grains and pseudopodia far extended.
C, Arcella mitrata, a transparent individual showing the protoplasmic body suspended within the shell; several vacuoles are shown, but no nucleus.
(Highly magnified.)

CHAPTER XIII.

UNICELLULAR ANIMALS (PROTOZOA) (*Continued*).

B Infusoria.
(*Paramœcium, Vorticella, etc.*)

INFUSORIA are minute unicellular animals found like *Amœba* in stagnant water or in organic infusions (see p. 201) (hence "Infusoria"). In the leading features of their organization they are closely similar to *Amœba* and its allies, from which they differ, however, in having a much higher degree of differentiation, in moving by means of cilia instead of pseudopodia, and in showing the first indication of gamogenesis (amphimixis). *Paramœcium* (the slipper-animalcule) is an actively free-swimming form often found in multitudes in hay-infusion or water containing the decomposing remains of *Nitella* and other water-plants. *Vorticella* (the "bell-animalcule") is commonly attached by a slender stalk to duck-weed (*Lemna*) and other water-plants, or to other submerged objects; at other times it breaks loose from the stalk and swims for a while actively about. The two forms are constructed upon essentially the same plan, but *Vorticella* shows in some respects a much higher degree of differentiation.

Paramœcium.—The slipper-shaped body (Fig. 87) is covered with cilia by means of which the animal rapidly swims about. Morphologically the body is a single cell, having the same general composition as in *Amœba*, but possessing in addition a delicate surrounding membrane ("cuticle") or cell-wall. The differentiation of the protoplasm into ectoplasm and entoplasm is very sharply marked, and the former contains numerous peculiar rod-like bodies (*trichocysts*) from which long threads may be thrown out. Their function is probably that of offence and protection. As in *Amœba* the protoplasm contains *water-vacuoles* ($w.v$) and *food-vacuoles* ($f.v$) (both of which are carried

THE SLIPPER-ANIMALCULE. 169

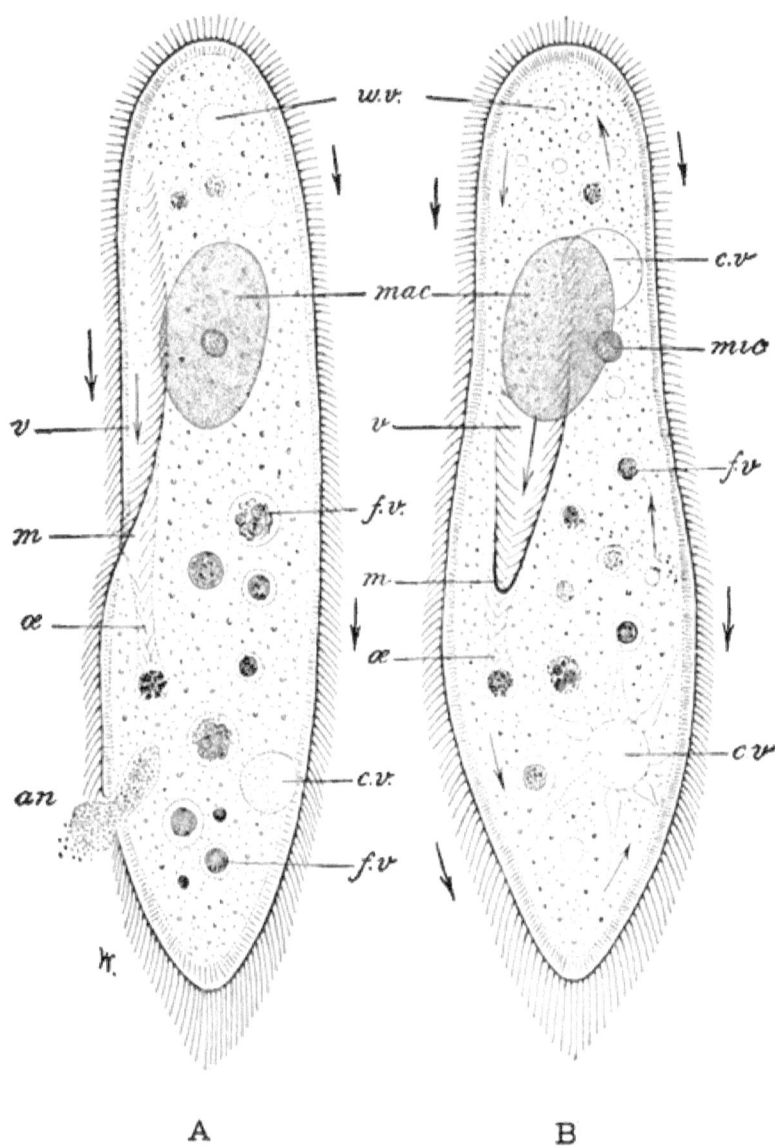

FIG. 87.—*Paramœcium caudatum.* A, from the left side, showing the anal spot; B, from the ventral side, showing the vestibule *en face*; arrows inside the body indicate the direction of protoplasmic currents, those outside the direction of water-currents caused by the cilia.

an, anal spot; *c.v.* contractile vacuoles; *f.v.* food-vacuoles; *w.v.* water vacuoles; *m.* mouth; *mac*, macronucleus; *mic*, micronucleus; *œ*, œsophagus; *v*, vestibule. The anterior end is directed upwards.

about by currents in the entoplasm), and two very large *contractile vacuoles* (*c.v*) occupying a constant position, one near either end of the body. The nucleus (as in Infusoria generally) is differentiated into two distinct parts, viz., a large oval *macronucleus* (*mac.*) and a much smaller spherical *micronucleus* (*mic.*) (double in some species) lying close beside it.

Unlike *Amœba*, *Paramœcium* possesses a distinct *mouth* (*m*) and *œsophagus* (*œ*) which open to the exterior through an oblique funnel-shaped depression known as the *vestibule* (*v*) situated at one side of the body. Minute floating food-particles are drawn by the cilia into the mouth and accumulate in a ciliary vortex at the bottom of the œsophagus. From time to time a bolus or food-mass is thence passed bodily into the substance of the entoplasm, forming a food-vacuole within which digestion takes place. The indigestible remnants are finally passed out not through a permanent opening or anus, but by breaking through the protoplasm at a definite point, hence known as the *anal spot*, which is situated near the hinder end (Fig. 87). The *contractile vacuoles* of *Paramœcium* are especially favorable for study, showing at the moment of contraction, or just before it, a pronounced star-shape, with long canals running out into the protoplasm. Through these liquid is supposed to flow into the vacuole.

Like *Amœba*, *Paramœcium* occurs both in an *active* and in an *encysted* state. In the former state it multiplies by transverse fission, division of both *macronucleus* and *micronucleus* preceding or accompanying that of the protoplasmic body (Fig. 88, *A*). Under favorable conditions division may take place once in twenty-four hours, or even oftener. This process, which is a typical case of agamogenesis, may be repeated again and again throughout a long period. But it appears from the celebrated researches of Maupas that even under the most favorable conditions of food and temperature the process has a limit (in the case of *Stylonichia*, a form related to *Paramœcium*, this limit is reached after about 300 successive fissions). As this limit is approached the animals become dwarfed, show various signs of degeneracy, and finally become incapable of taking food. The race grows old and dies.

In nature, however, this limit is probably seldom if ever

reached, and the degenerative tendency seems to be checked by a process known as *conjugation*. In this process two individuals place themselves side by side, partially fuse together, and remain thus united for several hours (Figs. 88, *B*, *C*). During this union an exchange of nuclear material is effected, after which the animals separate, both *macronucleus* and *micronucleus* now

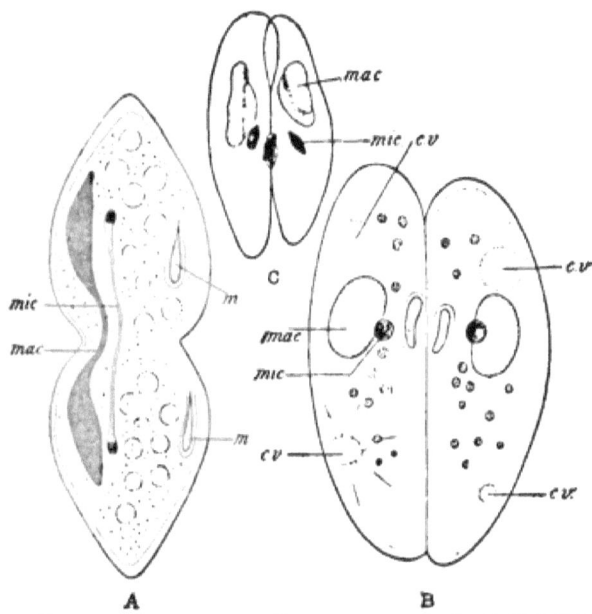

FIG. 88.—*A*. Fission of *Paramœcium*. (From a preparation by G. N. Calkins). *mac*, macronucleus; *mic*, micronucleus; *m*, mouth.
B. First stage of conjugation. The animals are applied by their ventral surfaces; the only change thus far is the enlargement of the micronuclei.
C. Conjugation at the moment of exchange of the micronuclei (less magnified). The macronuclei are degenerating. Each individual contains two micronuclei (now spindle-shaped), one of which remains in the body, while the other crosses over to fuse with the fixed micronucleus of the other individual (After Maupas.)

consisting of mixed material derived equally from both individuals. Separation of the two animals is quickly followed by fission in each.

In each individual the macronucleus breaks up and disappears. The micronucleus of each divides twice, and of the four bodies thus produced three disappear. The fourth divides again into two, one of which remains in the body, while the other crosses over and fuses with one of the micronuclei of the other individual, after which the animals separate. This process being reciprocal, each individual now contains a micronucleus containing an equal amount of material from each individual. This micronucleus now divides twice and gives rise to four bodies, two of which become macronuclei and two micronuclei. Fission next occurs, and is thereafter continued in the usual manner.

This is a process clearly analogous to the union of the germ-cells of higher animals. It cannot, however, be called gamogenesis or even reproduction; it is only comparable with one of the elements of gamogenesis. In the metazoön a fusion of two

FIG. 89.—Group of *Vorticella*, in various attitudes, attached to the surface of a water-plant.

cells (fertilization) is followed by a long series of cell-divisions (cleavage of the ovum), the resulting cells being associated to form one new individual. In the Infusoria temporary fusion (conjugation) is likewise followed by a series of cell-divisions, but the cells become entirely separate, each being an individual.

Vorticella agrees with *Paramoecium* in general structure, but differs in many interesting details, most of which are the expres-

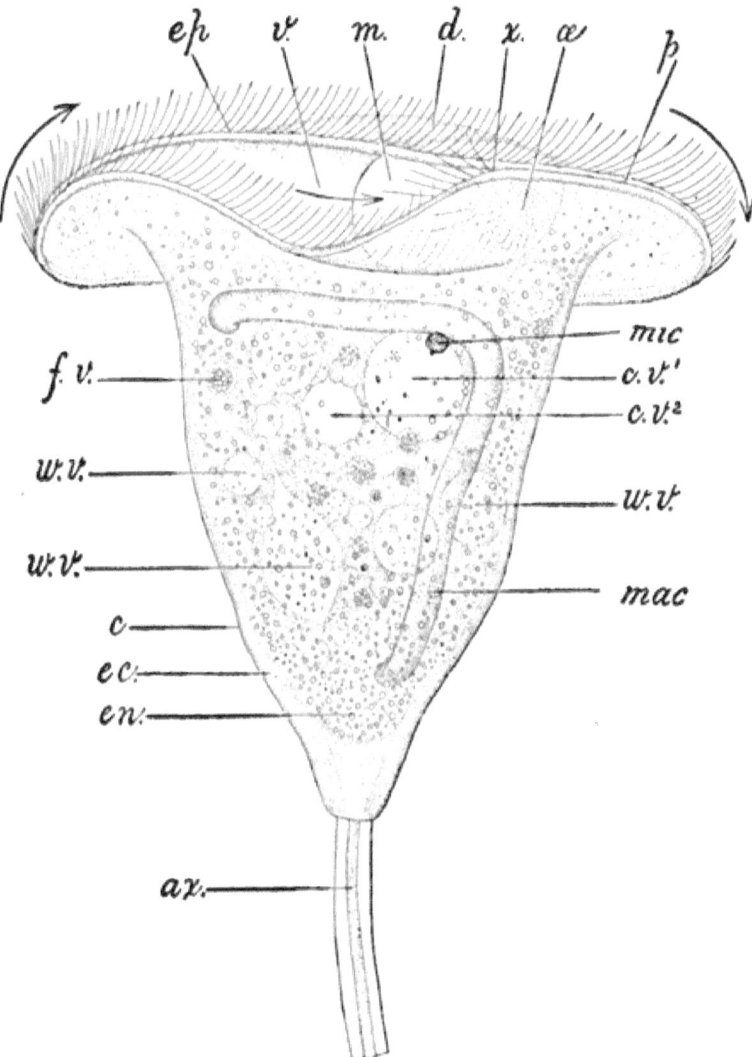

FIG. 90.—A single head of *Vorticella*, highly magnified. *cx*, contractile axis of the stalk; *c*, cuticle; *c.v*, contractile vacuole; *d*, disk; *ec*, ectoplasm; *en*, entoplasm; *ep*, epistome; *f.v*, food-vacuole; *m*, mouth; *mac*, macronucleus; *mic*, micronucleus; *œ*, œsophagus; *p*, peristome; *v*, vestibule; *w.v*, water-vacuoles; *x*, point at which epistome and peristome meet at one end of the vestibule.

sion of higher differentiation. The body is pear-shaped or conical, attached at its apex by a long slender stalk. The latter consists of a slender contractile *axial filament*, by means of which the stalk may be thrown into a spiral and the body drawn down, and an elastic *sheath* (continuous with the general cuticle) by which the stalk is straightened (Fig. 90). The cilia are confined to a thickened rim, the *peristome* (*p*), surrounding the base of the cone, which may be termed the *disk*. At one side the disk is raised, forming a projecting angle covered with cilia, and known as the *epistome* (*ep*). At the same side the peristome dips downwards, leaving a space between it and the epistome. This space is the *vestibule* (*v*), and into it the mouth opens. In it likewise is situated an *anal spot* like that of *Paramæcium*. The cilia produce a powerful vortex centering in the mouth, by means of which food is secured. The macronucleus (*mac*) is long, slender, and horseshoe-shaped; the small spherical micronucleus (*mic*) lies near its middle portion. There is usually but one contractile vacuole.

Vorticella multiplies by fission, division of the protoplasm being accompanied by that of the macronucleus and micronucleus (Fig. 91). The plane of fission is vertical (thus dividing the peristome into halves), but extends only through the main body, leaving the stalk undivided. At the close of the process, therefore, the stalk bears two heads. One of these remains attached to the original stalk, while the other folds in its peristome, acquires a second belt of cilia around its middle (Fig. 91), breaks loose from the stem, and swims actively about as the so-called "motile form." Ultimately it attaches itself by the base, loses its second belt of cilia, develops a stalk, and assumes the ordinary form. By this process dispersal of the species is ensured. Under unfavorable conditions similar motile forms are often produced without previous fission, the head simply acquiring a second belt of cilia, dropping off, and swimming away to seek more favorable surroundings. *Vorticella* may become encysted, losing its peristome and mouth, becoming rounded in form, acquiring a thick membrane, and having no stalk. In this state it is said sometimes to multiply by *endogenous division*, breaking up into a considerable number of minute rounded bodies (*spores*) each of which contains a fragment of the nucleus. These are finally liberated by the bursting of the membrane, acquire a ciliated belt, and after swimming for a time become attached, lose the ciliated belt, and develop a stalk and peristome.

Vorticella goes through a process of conjugation which has some interesting peculiarities. (1) Conjugation always takes place between a large attached individual (the *macrogamete*) and a much smaller free-swimming individual (the *microgamete*)

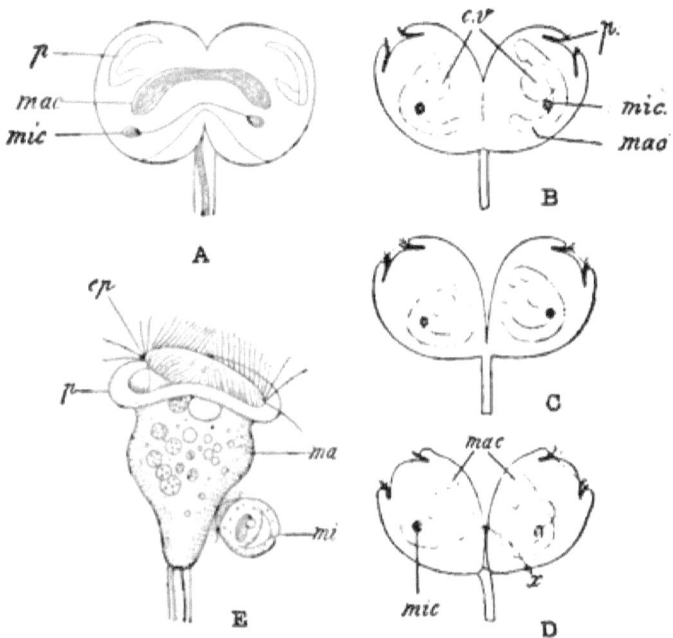

Fig. 91.—Fission and conjugation of *Vorticella*. *A*. Early stage of fission, showing division of micronucleus (*mic*) and macronucleus (*mac*); *p*, peristome. (After Bütschli.)

B, C, D. Successive stages of fission; in *B* and *C* the nuclei have completely divided and fission of the cell-body is in progress; *c.v*, contractile vacuoles. In *D* fission is complete; the right-hand individual has acquired a belt of locomotor cilia at *x*, and is ready to swim away.

E. Conjugation of a fixed macrogamete (*ma*) with a free-swimming microgamete (*mi*); *p*, peristome, *ep*, epistome. (After Greeff.)

(Fig. 91, *E*). The microgamete is formed either by the unequal fission of an ordinary individual, the smaller moiety being set free, or by two or more rapidly succeeding fissions of an ordinary individual. (2) Conjugation is permanent and complete, the body of the microgamete being wholly absorbed into that of the macrogamete. Within the body of the latter, after complicated changes, the nuclei fuse together, and this is followed by fission. The analogy of conjugation to the fertilization of the egg is here complete. The conjugating cells show a sexual differentiation, one being like the ovum, large and fixed, the other like the spermatozoön, small and motile.

As in *Paramaecium* the macronuclei entirely disappear, fusion takes place between derivatives of the micronuclei, and from the resulting body both macronuclei and micronuclei are derived.

Euglena and Other Simpler Infusoria. Besides forms like *Paramaecium* and *Vorticella* which bear numerous cilia, there are many Infusoria which possess only one large lash or *flagellum*. Of these *Euglena*, which is sometimes found in stagnant water, sewage-polluted pools, etc., is one of the most interesting, inasmuch as it contains chlorophyll, possesses an "eye-spot" of red pigment, and under certain conditions exhibits amœbiform movements.

Compound or "Colonial" Forms. In a number of forms, closely related to *Vorticella*, the individuals ("zooids") formed by fission do not immediately separate, but remain for a time united to form a "colony" which may contain hundreds of zooids. *Zoöthamnion*, a common species, thus forms a beautiful tree-like organism, consisting of a single central stalk with numerous branching offshoots from its summit, each twig terminating in a zooid. The entire system of branches is traversed by a continuous contractile axis. *Carchesium* is similar, but the axis is interrupted at the beginning of each branch. In *Epistylis* the entire axis is non-contractile.

Such colonial forms are of high interest as indicating the manner in which true multicellular forms may have arisen. From the latter, however, they differ not only in the fact that the association of the cells is not permanent, but in the absence of any division of labor among the units.

Physiology. Most Infusoria are true animals, agreeing with *Amœba* in the essential features of their nutrition, and having the power to digest not only proteids, but also carbohydrates and fats. *Paramaecium* and *Vorticella* are herbivorous forms, feeding upon minute plants, and especially upon the bacteria-

Other forms are omnivorous (e.g., *Stentor, Bursaria*), feeding both on vegetable and on animal food. Others still are carnivorous and lead a predatory life, often attacking herbivorous forms much larger than themselves, precisely as is the case with carnivores among the mammalia. Thus the unicellular world reproduces in miniature the essential biological relations of higher types.

It is a remarkable fact that some species of Infusoria (e.g., *Paramœcium bursaria, Vorticella viridis*) contain numerous chlorophyll-bodies embedded in the entoplasm. Much discussion has arisen as to whether these bodies are to be regarded as an integral part of the animal, i.e., differentiated out of its own protoplasm, or as minute plants living "symbiotically" (i.e. as mess-mates) within the animal. In the former case (which is the most probable) the animal would to a certain extent be nourished after the fashion of a green plant (cf. p. 148).

It will now be clear to any one who has carefully considered the phenomena described in the foregoing pages that the unicellular animals are "organisms" by right, and not merely by courtesy. In some of the Infusoria, for example, differentiation within the single cell may go so far as to give rise to primitive sense-organs (as in the case of the eye-spot of *Euglena*); a rudimentary œsophagus and definite mouth (as in *Paramœcium* and *Vorticella*); organs of locomotion (*cilia, flagella*); organs of excretion (contractile vacuoles) etc., etc.

CHAPTER XIV.

UNICELLULAR PLANTS.

A. Protococcus.

(*Protococcus, Pleurococcus, Chlorococcus, Hæmatococcus, etc.*)

UNICELLULAR plants, like unicellular animals, are very common, although as individuals mostly invisible on account of their microscopic size. In the mass, however, they are often visible either as suspended or floating matter, causing "turbidity" in liquids (*yeast, bacteria, diatoms, desmids,* etc.) or discolorations on tree-trunks, earth, stones, roofs, and flower-pots. (*Protococcus, Glaeocapsa,* etc.).

Under the term *Protococcus* ($\pi\rho\omega\tau os$, *first*, $\kappa\acute{o}\kappa\kappa o\varsigma$, *berry*) we may for our present purposes include a number of the simplest spherical forms, generally green in color and of uncertain affinities in classification, but very similar in structure, living for the most part in quiet waters or on moist earth, stones, tree-trunks, or old roofs, or in water-butts, roof-gutters, and the like. Sometimes the color which they exhibit is yellowish-green, sometimes bluish-green, and sometimes, though less often, reddish, according to the species.

One of the commonest and most conspicuous is a species often seen on the shady side of old tree-trunks where, when abundant, it forms a greenish dust-like coating or discoloration, scarcely visible when dry but becoming a rich bright green during prolonged rains or after warm showers. If pieces of bark covered with this form of *Protococcus* are moistened, the greenish coating may be observed at any time. It is granular in texture and after moistening is easily loosened by a camel's-hair brush.

Morphology. Microscopical examination shows that the particles detached consist of rounded yellowish-green cells occurring either singly or in groups of two, three, four, or even more.

Each single cell is a complete individual, capable of carrying on an independent life. It fairly represents the green plant (such as *Pteris*) reduced to its lowest terms. (Fig. 92.)

Like *Amaba* and the Infusoria *Protococcus*, at least in some species, occurs both in a *motile* or active state in which it moves about, and a quiescent or *non-motile* state analogous to the encysted state of the unicellular animals. In the latter the motile or active state is the usual or dominant condition and the encysted state is rarely assumed. In *Protococcus*, on the other hand, the motile state is rare, and the ordinary activities of the plant are carried on in the non-motile state.

Structure. In structure *Protococcus* is a nearly typical cell (p. 22). It consists essentially of an approximately spherical mass of protoplasm enclosed within a thin woody layer of cellulose (cell-wall or cell-membrane), and contains a single nucleus. It also includes one or more *chlorophyll-bodies* (*chromatophores*) (p. 126) by virtue of which it is able to manufacture its own foods, very much after the fashion of the green cells of *Pteris*.

In those forms which possess a motile stage the latter consists of a spherical, egg-shaped or pear-shaped cell having chromatophores and a membrane through which two flagella protrude. In the oval forms these are placed near the narrowed end of the cell, and in all cases they are locomotor organs and propel the cell swiftly through the water. (Fig. 92).

Reproduction. The ordinary method of reproduction in the unicellular plants, as in the unicellular animals, is by cell-division. In *Protococcus* the sphere becomes divided by a partition into two cells which eventually separate completely one from the other. Very often, however, the separation being incomplete or postponed until after each daughter-cell has in turn become divided, groups or aggregates of cells arise which suggest the first steps in the formation of tissue in the development of higher forms. In the end, however, separation is total and complete, and each cell is therefore not a unit in a body, but is itself a body and an individual (see p. 156). (Fig. 92.)

The daughter-cells thus produced are the young, or offspring, which have the power to grow and ultimately to divide in their turn. Under favorable circumstances generation may thus follow generation in quick succession. Each young cell is actually

180 UNICELLULAR PLANTS.

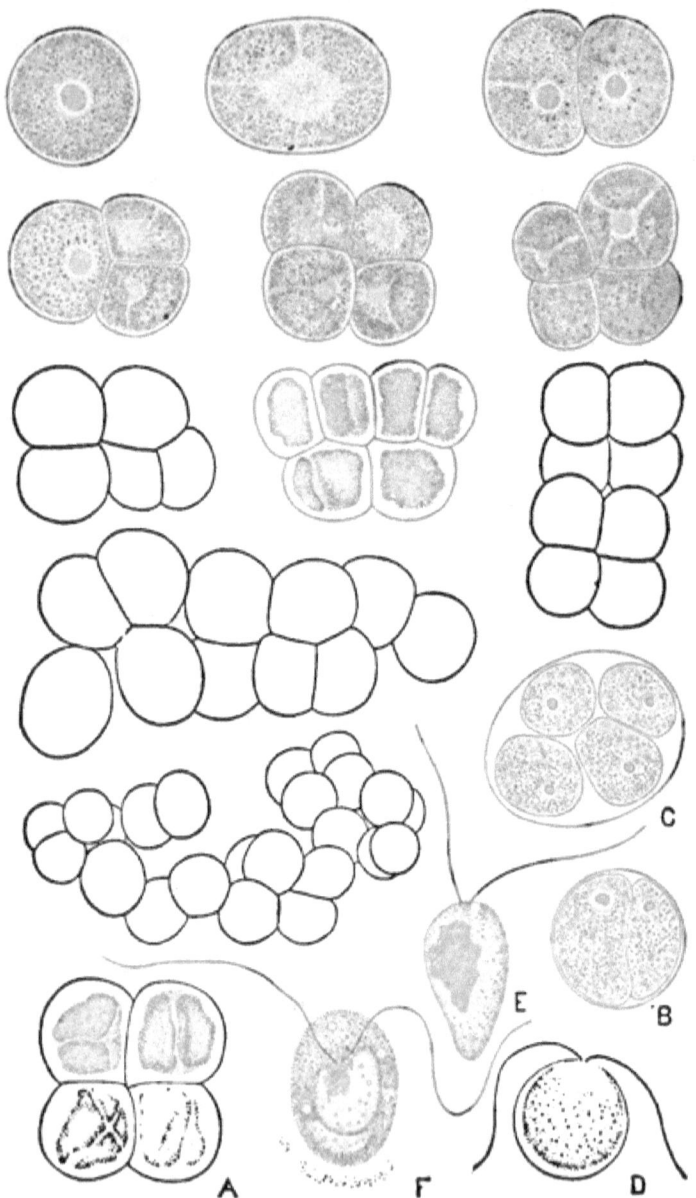

FIG. 92.—Protococcus (*Pleurococcus*) from the bark of an elm tree, in active vegetation and showing aggregation into masses of cells. *A*, *Pleurococcus* in the dried condition. *B*, *Ascococcus* (?), showing endogenous division into two cells and (*C*) into four. *D*, *E*, *F*, motile forms of *Protococcus* (after Cohn).

one half of the parent cell and contains a moiety of whatever that contained. Here, therefore, as in *Amœba*, the problems of heredity, uncomplicated by the occurrence of sex, are reduced to their lowest terms.

In some kinds of *Protococcus* the quiescent cells, under special circumstances, which are not well understood, give rise to the *motile* forms (*zoospores*) referred to above. Cilia, or rather flagella, are formed, and the protoplasmic mass with its included chromatophores swims actively about in the water. After a time these motile cells may come to rest, lose their flagella and divide into two or more daughter-cells, each of which in its turn may become a motile cell and repeat the process, or, under other conditions, develop into the ordinary quiescent cell.

In some species of *Protococcus* in which there is a motile stage another form of reproduction, a kind of rudimentary gamogenesis, has been observed. In this process two of the motile cells (gametes) meet, fuse (*conjugation*), lose their flagella, become encysted (see p. 161), and ultimately give rise to the ordinary cells of *Protococcus*, both non-motile and motile. This process, however, has not yet been observed in the species under consideration.

Physiology. Our actual knowledge of the physiology of *Protococcus* is very small. But the study of comparative plant physiology gives every reason to believe that the essential physiological operations of this simple plant are fundamentally of the same character as in the higher green plants, such as *Pteris*.

Nutrition. The income of *Protococcus*, when growing in its natural habitat on tree-branches, moist bricks, and the like, is difficult to determine. But as it is able to live also in ordinary rain-water, we are able to set down its probable income under those conditions with some degree of accuracy. There is do doubt that it absorbs water and carbon dioxide by diffusion through the cellulose wall, and that these substances are used in the manufacture of starch, which, if stored up, makes its appearance in the form of small granules within the chromatophores. This process takes place only in the light and through the agency of the chlorophyll, and is attended by a setting free of oxygen precisely as in *Pteris*. Nitrogen is probably derived from nitrates or ammoniacal compounds, minute quantities of which are dissolved in the water, and other necessary salts (sulphates, chlorides, phosphates, etc.) as well as free oxygen are procured from the same source. These substances may be derived from dust blown or washed by the rain into the water, or from the walls of the vessel. To the process of starch-making, attended by the absorption of CO_2 and H_2O and the liberation of O, the term "assimilation" is generally given. Like other plants, moreover, *Protococcus* probably breathes by absorbing free oxygen and setting free CO_2 (respiration).

The income and outgo of *Protococcus* may then be displayed by the following diagram:

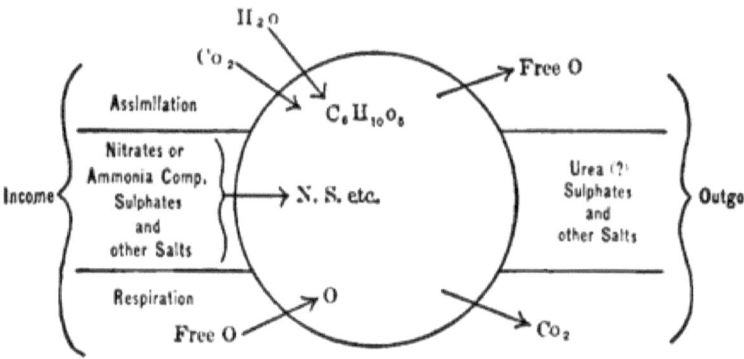

It should be understood that this only represents the broad outlines of the process and under the simplest conditions. It is quite possible that under other conditions *Protococcus* may use more complex foods. The facts remain, however, (1) that *Protococcus* is dependent on the energy of light; (2) that its action is on the whole constructive, resulting in the formation of complex compounds (carbohydrates, proteids) out of simpler ones. In these respects it shows a complete contrast to *Amœba*, which is on the whole destructive, breaking down complex compounds into simpler ones, and is independent of light, since it derives energy from the potential energy of its food. The relations between *Protococcus* and *Amœba* are therefore an epitome of the relations between *Pteris* and *Lumbricus*, and between green plants and animals generally.

The Fundamental Physiological Properties of Plants. In considering the physiology of *Amœba* we found it possible to reduce its vital activities to a few fundamental physiological properties, namely, contractility, irritability, metabolism, growth and reproduction, common to all animals. A little reflection will show that the same properties are manifested also by *Protococcus*. Contraction and irritability are difficult to witness in the quiescent stage of *Protococcus*, but obvious enough in the rarer motile forms. Metabolism, growth and reproduction, on the other hand, are evident accompaniments of normal life, even in the quiescent condition. And precisely as *Protococcus* differs from *Amœba* in respect to contractility and irritability, of which it possesses relatively little, so plants in general differ in these respects from animals in general. Animals are eminently contractile and irritable, while plants are but feebly specialized in these directions. On the other hand, as we have already seen in comparing *Pteris* with *Lumbricus* (p. 154), and as we see once more in comparing *Protococcus* with *Amœba*, in respect to metabolism, the green plant is pre-eminently constructive, while the animal is preëminently destructive, of organic matter.

In their modes of nutrition, as stated above, *Amœba* and *Protococcus* represent two physiological extremes. We pass now to the study of Yeasts and Bacteria, which are plants *destitute of chlorophyll* and in a certain sense may be regarded as occupying a middle ground between these extremes.

Other Forms. There are innumerable species of unicellular green plants. A vast group of peculiar brownish forms covered with transparent glass-like cells composed of siliceous material is known as the *Diatomaceæ* or *diatoms*. In these the chlorophyll is masked by a brown pigment, but is nevertheless present. Another group is that known as the *Desmidiæ* or *desmids*. These often have the individual cells peculiarly constricted in the middle so that at first sight the two halves appear to be two separate cells. More closely resembling *Protococcus* in many respects are some members of the *Cyanophyceæ* or "blue-green algæ," among which *Chroöcoccus* and *Glœocapsa* differ from *Protococcus* chiefly, in the former case, in having a blue-green instead of a yellow-green pigment, and, in the latter, not only in this respect, but also in the fact that the single cells are widely separated by transparent mucilage.

CHAPTER XV.

UNICELLULAR PLANTS (*Continued*).

B. Yeast.

(*Saccharomyces.*)

UNDER the general name of *yeast* are included some of the simplest forms of vegetal life. Some yeasts are "wild," living upon fermenting fruits or in fruit juices, and commonly

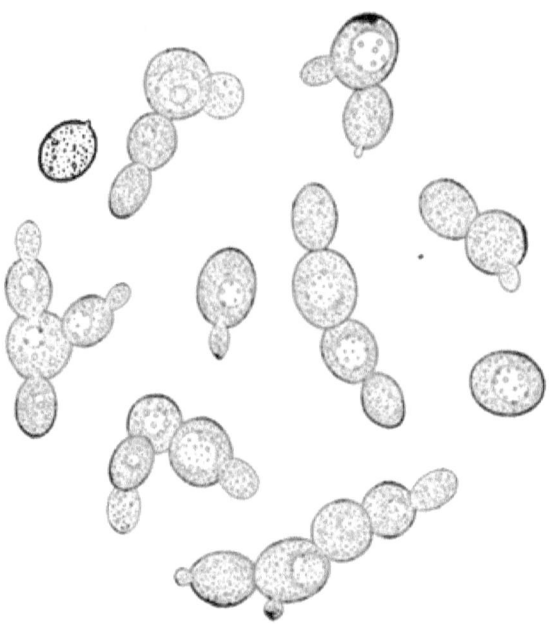

FIG. 93.—Yeast-cells. Brewer's (top) yeast actively vegetating. The large internal vacuoles and the small fat-drops are shown, as are also buds, in various stages of development, and the cell-wall. Nuclei not visible. (Highly magnified.)

occurring in the air; others are "domesticated," or cultivated, such as those regularly employed in brewing and in baking.

If a bit of "yeast-cake" (either "compressed" or "dried" yeast) is mixed with water, a milky fluid is obtained which closely resembles the so-called baker's or brewer's yeast.

Microscopical examination proves that the milky appearance of liquid yeasts is due chiefly to the presence of myriads of minute egg-shaped suspended bodies, and that pressed yeast is almost wholly a mass of similar forms. These are the cells of yeast; which is therefore essentially a mass of unicellular organisms. For reasons which will soon appear yeast is universally

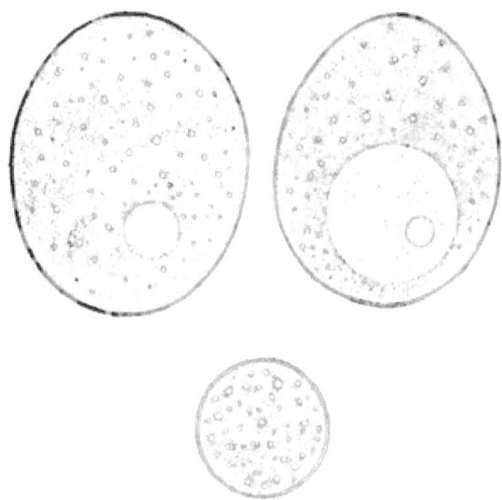

FIG. 94.—Yeast-cells. Brewer's (bottom) yeast showing structure—protoplasm, cell-walls, vacuoles, fat-drops. (Nuclei not shown.)

regarded as a plant, and the single cell is often spoken of as the yeast-plant.

Morphology. The particular yeasts which we shall consider are the common cultivated forms of commerce. The cells of an ordinary cake of pressed yeast are spherical, spheroidal, or egg-shaped in form, and consist of a mass of protoplasm enclosed within a well-defined cell-wall. By appropriate treatment the latter may be shown to consist of cellulose; and it is distinctly thicker in old or resting cells than in young ones or those vigorously growing. Within the granular protoplasm (*cytoplasm*) are usually a number of vacuoles (containing sap) and minute shining dots (probably fat-droplets), but

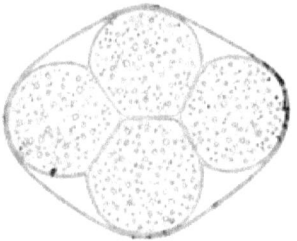

FIG. 95.—Spores of Yeast (Ascospores). Four spores in a cell of brewer's yeast *Saccharomyces cerevisiæ*.

no chlorophyll is present and no starch. Until recently the yeast-cell was supposed to be destitute of a nucleus, but it is now known that each cell probably possesses a large and characteristic nucleus. This, however, can be demonstrated only by special reagents and is rarely or never seen in the living cell (Fig. 96).

Reproduction. The ordinary mode of reproduction of yeast is by a modification of cell-division called *budding*. Under

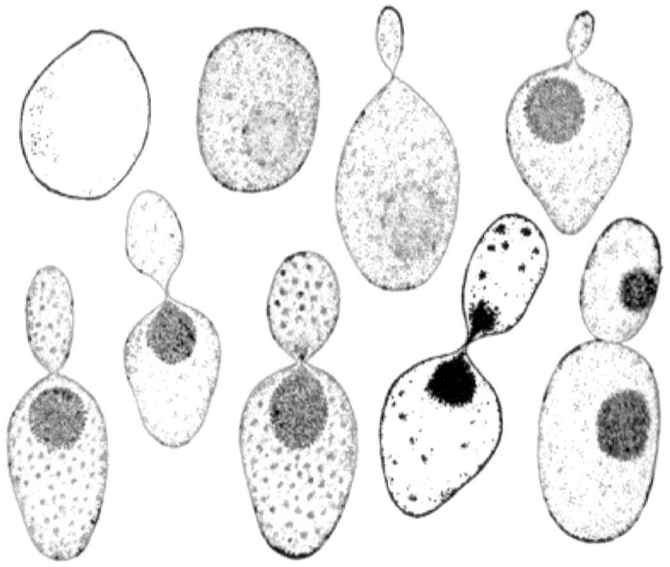

FIG. 96.—The Nuclei of Yeast-cells and the Process of Budding. (Drawn by J. H. Emerton from specimens prepared by S. C. Keith, Jr.) The upper left-hand figure shows the nucleus in a specimen treated with Delafield's hæmatoxylin. The other figures in the upper row and those in the lower (from left to right) show cells in successive stages of budding, together with the appearance, position, and movements of the nucleus. It will be observed that the bud is formed before the nucleus divides. (Iron-hæmatoxylin method.)

favorable circumstances in actively growing yeast a local bulging of the wall takes place, usually near, but not precisely at, one pole of the cell. Protoplasm presses into this dilatation or "bud" and extends it still further. At this time we have still but one cell, although it now consists of two unequal parts and the separation of a daughter-cell is clearly foreshadowed. Eventually the connection between the two parts is severed and the daughter-cell or "bud" is detached from the original or parent-cell; but detachment may or may not occur until after the bud has begun to produce daughter-cells in its turn, and more than one bud may be borne by either or both parent- or daughter-cells. In very rapid growth the connection may persist between the cells even during the formation of several generations of buds; but this is unusual, and in cases where a number of cells remain apparently united together forming tree-like forms there is often no real connection, the cells separating readily on agitation.

Endospores (Ascospores). Some yeasts in addition to the method of reproduction by budding exhibit another mode known

 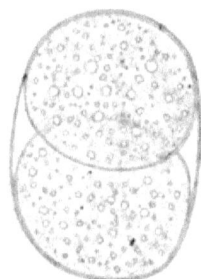

FIG. 97.—Spores of Yeast (Ascospores). Three- and two-celled stage of spore formation in *S. cerevisiæ*.

as *endogenous division* or *ascospore formation*. Under certain circumstances not yet entirely understood there are formed *within the yeast-cell* two, three, or four rounded shining spores. These become surrounded by thick walls and thus give rise eventually to a group of daughter-cells within the original cellulose sac. To the latter the term *ascus* (sac) has been applied, and to its contained daughter-cells the term *ascospores*.

It is not yet allowed by all botanists that this terminology, which implies a relationship of yeasts to the Ascomycetous fungi, is sound; but it is commonly used.

Each ascospore is capable under favorable circumstances of sprouting and starting a new series of generations of ordinary yeast-cells. It should be particularly observed that the endospores of yeast are reproductive bodies, and that the process of their formation is one of multiplication—not merely one of defence or protection, as is the case with the so-called "spores" of bacteria described beyond (p. 194).

Physiology. Like all other organisms the yeast-plant occupies a definite position in space and time; it possesses an environment with which it must be in harmony if it is to live, from which it derives an income, and to which it contributes an outgo of matter and energy; it manufactures its own substance from foods (*anabolism*), and like all living things it wastes by oxidation of its substance (*katabolism*). It is not obviously contractile or irritable, but it is highly metabolic and reproductive.

Yeast and its Environment. Yeast is an aquatic form, and, as might be supposed, cultivated yeast thrives best in its usual habitat, the juices of fruits, such as apples or grapes, and the watery extracts of sprouted seeds, such as barley, corn, and rye (wort, mash, etc.). It lives, however, more or less successfully in many other places (such as the dough of bread), and can even endure much dryness, as is shown by the commercial "dried-yeast." It appears to prefer a temperature from $20°$ to $30°$ C.; it is usually killed by boiling, but if dried, it can endure high temperatures. Its action is inhibited by very low temperatures, but like most living things it endures low temperatures better than high. It is killed by many poisons (antiseptics).

Income. Owing to its industrial importance yeast has been perhaps more thoroughly studied in respect to its nutrition than any other unicellular organism. And yet it is impossible to give accurate statistics of its normal income and outgo. It is believed that the ordinary income of a yeast-cell living in wort (the watery extract of sprouted barley-grains) consists of *a, dissolved oxygen; b, nitrogenous bodies* allied to proteids, but diffusible and able to pass through the cellulose wall; *c, carbohydrates, especially sugary matters*; and *d, salts* of various kinds.

It was supposed for a long time by Pasteur and others that yeast could dispense with free (dissolved) oxygen in its dietary. It now appears that this faculty is temporary only, and that if yeast is to thrive it must, like all other living things, be supplied, at least occasionally, with free oxygen.

Metabolism. Out of the income of foods just described yeast is able to build up its own peculiar protoplasm (*anabolism*), and, doubtless, to lay down the droplets of fat which often appear in it. There is good reason to believe that its substance also breaks down, with the production of carbon dioxide, water, and nitrogenous waste (*katabolism*), and the concomitant liberation of energy. The work to be done by the yeast-cell is plainly limited. The manufacture of new and of surplus protoplasm and the protrusion of buds require work, partly chemical, partly mechanical; but most of the liberated energy probably appears as heat. In point of fact, great activity of yeast is accompanied by a rise of temperature, as may be proved by placing a thermometer in "rising" dough or fermenting fruit-juice.

Outgo. Barring the outgo of energy already mentioned, and the probable excretion of carbon dioxide and nitrogenous waste, but little can be said concerning the outgo of a yeast-cell. The ordinary excretions are so masked by the presence of foreign matters in the liquids which yeast inhabits that little is known of the real course of events. To the consideration of conditions which entail these difficulties we may now pass, merely pausing to caution the student against the supposition that the evolution of carbon dioxide in fermentations represents to any great extent the normal respiration of the yeast cells.

Mineral Nutrients of Yeast. It has been shown (pp. 148, 181) that *Pteris* and *Protococcus*, inasmuch as they possess chlorophyll can live upon simple inorganic matters such as CO_2, H_2O, and nitrates, out of which they are able to manufacture for themselves energized foods such as starch. Yeast is unable to do this, as might be supposed from the fact that it is destitute of chlorophyll. And yet yeast does not require proteid ready-made as all true animals do, for experiments have shown that it can live and grow in a liquid containing only mineral matters plus some such compound of nitrogen as ammonium tartrate $(C_4H_4(NH_4)_2O_6)$. Upon a much less complex organic compound of nitrogen such as a nitrate it cannot thrive, thus showing its inferiority in constructive power to *Protococcus* and all green plants, on the one hand, and its superiority to *Amœba* and all animals, on the other.

Pasteur's fluid, composed of water and salts, among which is ammonium tartrate (above), will suffice to support yeast. It will support a much more vigorous growth if sugar be added to it. But if ammonium nitrate is substituted for ammonium tartrate yeast will refuse to grow in the fluid.

Yeast is a Plant. The superior constructive faculty of yeast, just described, separates it fundamentally from all animals in respect to its physiology, and allies it closely to all plants. Its inferiority to the chlorophyll-bearing plants or parts of plants, on the other hand, in no wise separates it fundamentally from plants; for it must not be forgotten that the power, even of plant-cells to utilize mineral matters as raw materials and from them to manufacture foods like starch, ordinarily resides exclusively in the chlorophyll bodies, and is operative only in the presence of light. It follows, therefore, that most of the cells, even of the so-called green plants, and a considerable portion of the contents of the so-called green cells, must be destitute of this synthetic power. Considerations of this kind show how exceedingly localized and special the starch-making function is, even in the "green" plants; and yeast probably compares very favorably in its synthetic powers with many of the colorless cells of such plants, or even with the colorless protoplasmic portions of chromatophore-bearing cells.

But yeast is vegetal rather than animal, morphologically as well as physiologically. Its structure more nearly resembles that of some undoubted plants (fungi) than any animal. Its wall is composed of a variety of cellulose, called fungus-cellulose; and cellulose, though occasionally occurring in animal structures, is, broadly speaking, a vegetal compound. Finally, in its methods of reproduction by budding, and by spores, yeast is allied rather to plants than animals.

Top Yeast. Bottom Yeast. In the process of brewing two well-marked varieties of yeast occur, known as "top" and "bottom" yeast. The former is used in the making of English ale, stout, and porter; the latter in the making of German or "lager" beer. The top yeast is cultivated at the ordinary summer temperature of a room, without special attention to temperature; the latter in rooms artificially cooled so that even in summer, icicles often hang from the walls. The two yeasts also show obvious differences in form, size, and structure; and how much they must differ in their function is plain from the very different products to which they give rise.

Wild Yeasts. Besides the commercial or cultivated yeasts there are also wild yeasts, and to them are due in the main the fermentations of apple-juice, of grape-juice, and other fruit juices. A drop of sweet cider shows under the microscope a good example of one of these species; and Pasteur long ago proved that the outer skins of ripe grapes and other fruits are apt to harbor yeast-cells in the dust which lodges upon them. More recently it has been shown that wild yeasts often live under apple-trees upon the surface of the earth. In a dry time the wind easily lifts the dust containing them and conveys them over great distances (cf. *Amœba*, Infusoria, etc.). The domesticated yeasts of to-day are probably the descendants of similar wild yeasts.

Red Yeast. One of the finest of the wild yeasts is the so-called "red yeast," which is furthermore very easy to study. Red yeast, and many others not red, grow luxuriantly upon a jelly, made by thickening beer-wort with common gelatine. In this way "pure" cultures—that is, cultures free from other species of yeasts, or bacteria, and consisting of one kind only—can be easily made and studied. The microscope shows that the cells of red yeast, which form red dots upon such jelly, are not themselves colored, but the pigment appears to lie between the cells, as in the case of the "miracle germ" (*Bacillus prodigiosus*).

Fermentation. To the processes where yeast is employed to produce chemical changes in various domestic, agricultural, and industrial operations the term *fermentation*, or more often *alcoholic fermentation*, is applied. In the "raising" of bread or cake, in brewing, cider-making, etc., yeast acting upon sugar produces from it an abundance of alcohol and carbon dioxide. Both products are sought for in brewing, and carbon dioxide is especially desired in bread-making.

But alcoholic fermentation is only one example of a large class, and yeast is only one of many ferments. We may, therefore, postpone further consideration of fermentation to the next chapter.

Related Forms. It has been shown by the researches of Hansen that ordinary commercial yeast is seldom one single species, as was formerly supposed, but rather a mixture of several species. It is therefore no longer safe to speak of commercial yeast as *Saccharomyces cerevisia*, unless careful examination by the modern methods has shown it to be such; and to determine what species exist in any particular specimen is often a laborious and difficult matter.

Inasmuch as the natural position of yeast in the vegetal kingdom is not established beyond all doubt, it is impossible to state precisely what are its near relatives. There are numerous unicellular colorless plants, but they are not necessarily closely related to yeast; and the student must not conclude for plants any more than for animals that because an organism unicellular it is necessarily at the very bottom of the scale of life.

CHAPTER XVI.

UNICELLULAR PLANTS (*Continued*).

C. Bacteria.

(*Schizomycetes.*)

THE smallest, and the most numerous, of all living things are the bacteria. Bacteria occur almost everywhere: they are lifted into the atmosphere as dust particles, in it they float and with its currents they are driven about; water—both fresh and salt—often contains large numbers of them; and the upper layers of the soil teem with them. But they are most abundant in liquids containing dissolved organic matters, especially such as have stood for a time—for example, stale milk and sewage, these fluids often containing millions of individual bacteria in a single cubic centimetre.

In respect to their abundance in the surface layers of the earth (one gram of fertile soil often containing a million or more), and the work which they do there in producing the oxidation of organic matters and changes in the composition of the soil, bacteria may well be compared with earthworms (cf. p. 42). They are also of much general interest because some are what are known as "disease-germs." Most bacteria, however, are not *parasitic*, but *saprophytic*, i.e., live upon dead organic matters, and therefore are not merely harmless, but positively useful in rendering back to the inorganic world useless organic matters. Some species such as the vinegar bacteria are commercially important.

In systematic botany bacteria constitute a well-defined group, the *Schizomycetes* (*fission-fungi*), their near allies being the *Cyanophyceæ* or "blue-green algae."

Morphology. Under the microscope bacteria appear as minute rods (*Bacilli*) (Fig. 98), balls (*Cocci*) (Fig. 100), or spirals (*Spirilla*) (Fig. 104), sometimes at rest, but often, at least in the case of the rods and spirals, in active motion. Little or no

SHAPES OF BACTERIA.

structure can be made out in them by the beginner, to whom they usually appear at first sight like pale, translucent or watery bits of protoplasm. Investigation has shown, however, that they possess a cell-wall (probably composed of cellulose) and a non-homogeneous protoplasm. Unlike *Protococcus*, but like yeast-cells, the cells of bacteria contain no chlorophyll. Nuclear mat-

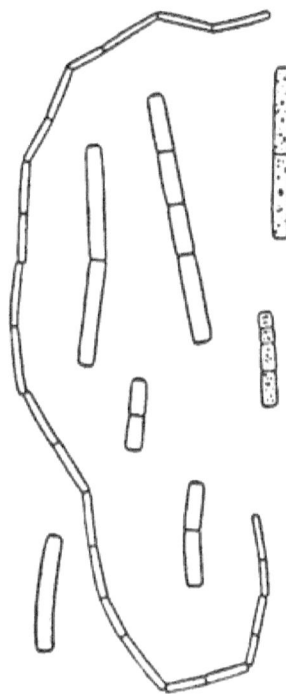

FIG. 98. — Bacillus Megaterium Rods (unstained) in various aggregations as commonly seen with a high power after their cultivation in bouillon and while rapidly growing and multiplying by transverse division.

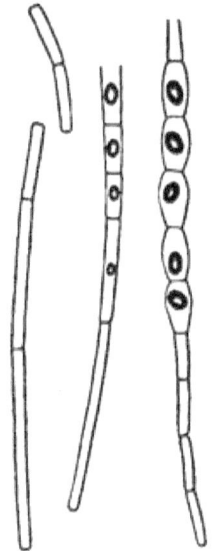

FIG. 99. — Bacilli from Hay Infusion (unstained). The filaments at the left in a condition of active vegetation. The middle filament forming spores. The filament to the right contains five spores enclosed in otherwise empty cells, the walls of which bulge, probably from the absorption of water.

ter is present, either scattered about, or, if the views of Bütschli be accepted, composing most of the protoplasmic body itself. Many bacteria bear appendages in the shape of flagella or cilia; but these can only be demonstrated in special cases, and by special methods. They are believed to be locomotor organs, and in some cases have been seen in active motion (Fig. 103).

The minuteness of bacteria is extraordinary. Many bacilli are not more than .005 mm. ($\frac{1}{5000}$ inch) in length or more than .001 mm. ($\frac{1}{25000}$ inch) in breadth. Some are very much smaller.

Most bacteria are at some time free forms; but like other unicellular organisms many of them have the power to pass from a free-swimming (*swarming*) into a quiescent (*resting*) condition. In the latter some undergo a peculiar change, in which the cell-wall becomes mucilaginous, and by the aggregation of numerous individuals or by repeated division lumps of jelly-like consistency (*zoöglœa*) arise. If the jelly mass takes the shape of a sheet or membranous skin (as happens in the mother-of-vinegar), it is sometimes described as *Mycoderma* (*fungus-skin*) (Fig. 102).

Reproduction. The bacteria increase in numbers solely by transverse division. Growth takes place and is followed by transverse division of the original cell, usually into halves. Each half then likewise grows and divides in its turn. In this way multiplication may go on in geometrical progression, and with almost incredible rapidity. It has been stated that such repeated divisions may follow only an hour apart, and on this basis it is easy to compute the enormous numbers to which a single cell may give rise in a single day.

If separation after division is complete, strictly unicellular forms arise. If actual separation is postponed, long rods, chains, or plates (in the case of cocci) may appear. Different names are given to the resulting forms. *Streptococcus* is a moniliform or necklace-like arrangement; *Staphylococcus*, single cocci; *Diplococcus*, cocci in pairs; *Leptothrix*, a filament of bacilli; *Sarcina*, a plate of cocci resembling a card of biscuit, or two or more cards superposed; etc., etc.

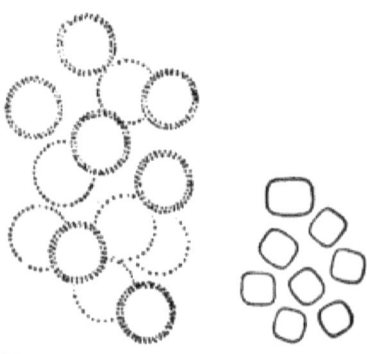

FIG. 100.—Micrococci (unstained) from hay infusion.
FIG. 101.—Short Bacilli (unstained) from hay infusion.

Spores. Some bacteria produce so-called *spores* (*endospores*) in the following way: The contents of the cell withdraw from the wall and condense into a (usually oval) mass at one end of the cell, leaving the rest of it empty It is at this time that the cell-wall is best seen. The condensed mass now becomes dark and opaque, apparently from the deposit upon itself of a greatly thickened and peculiar wall; it refuses to absorb stains which the original cell would have taken, and becomes exceedingly resistant to extremes of heat, cold, and dryness (Fig. 105). To these spores the Germans give the excellent term *Dauersporen*, i.e., *enduring spores*, often called *resting* spores.

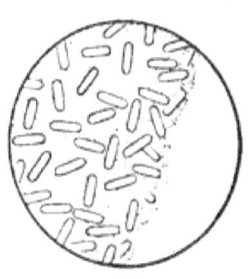

FIG. 102.—The Mother-of-Vinegar. The edge of a film of zoogloea of mother-of-vinegar as it appears under a high power. The bacteria are seen imbedded in the jelly which they have secreted.

When brought under favorable conditions, these sprout and, the ordinary bacterium cell having been produced, growth and fission proceed as before. Obviously these spores are very different in function from those of *Pteris* (p. 130), since they are protective merely, and not reproductive. They correspond, doubtless, to that phase of animal life which is known as the "encysted" state. Another mode of spore-formation in bacteria is that known as the production of *arthrospores*, in which a long slender cell may become constricted and detach daughter-cells from one or both ends. This is obviously a special case of unequal cell-division, but if it exists at all (which has been doubted) it clearly approaches agamogenesis in such forms as *Pteris*.

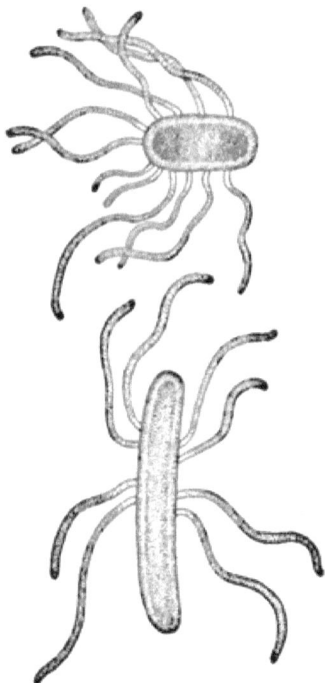

FIG. 103.—Ciliated Bacteria. The bacillus of typhoid fever, showing cilia. (From a specimen prepared by S. C. Keith, Jr. Drawn by J. H. Emerton.)

Physiology. Income, Metabolism, and Outgo. The bacteria

show a surprising diversity in the precise conditions of their nutrition, and it is therefore difficult to make for them a satisfactory general statement. As a group, however, and disregarding for the moment certain important exceptions, they are to be regarded as colorless plants living for the most part upon complex organic compounds from which they derive their income of matter and energy and which they decompose into simpler compounds poorer in potential energy. In so doing they bring about certain chemical changes in the substances upon which they act which are of the highest theoretical interest, and sometimes of great practical importance. Perhaps the most peculiar feature of the physiology of bacteria is the fact that while they are themselves individually invisible, they collectively produce very conspicuous and important changes in their environment. For example, vinegar bacteria act upon alcohol (in cider, etc.) and by a process of oxidation slowly convert it into acetic acid and water, thus:—

Fig. 104.—Spirillum undula. Spiral bacteria deeply stained. Drawn from the first photographic representation of bacteria ever published, viz., that of Robert Koch, in Cohn's *Beiträge*, 1876.

$$C_2H_6O + O_2 = C_2H_4O_2 + H_2O.$$

Here it is not the bacteria that are most conspicuous, but the effect which they produce. It is clear that the alcohol can be only one factor in the nutriment of the organism, because it contains no nitrogen, and the above reaction cannot represent more than a phase in the nutrition of the bacterium. That this is indeed the case is proved by the fact that if the conditions be somewhat changed the same bacteria may go further and convert the acetic acid itself into carbonic acid and water:—

$$C_2H_4O_2 + O_2 = 2CO_2 + 2H_2O.$$

Chemical changes of this kind in which the effect upon the environment is more conspicuous than, and out of all proportion to, the change in the agent are in some cases known as *fermentations*, and the agent effecting the change is described as a *ferment*. Some ferments are *organized* or *living*, and some are

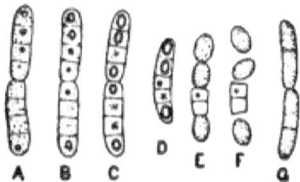

FIG. 106.—Bacillus megaterium (× 600). Spore formation and germination. *A*, a pair of rods forming spores, about 2 o'clock P.M. *B*, the same about an hour later. *C*, one hour later still. The spores in *C* were mature by evening; the one apparently begun in the third upper cell of *A* and *B* disappeared; the cells in *C* which did not contain spores were dead by 9 P.M. *D*, a five-celled rod with three ripe spores, placed in a nutrient solution, after drying for several days, at 12.30, P.M. *E*, the same specimen about 1.30 P.M. *F*, the same about 4 P.M. *G*, a pair of ordinary rods in active vegetation and motion. (After De Bary.)

unorganized or *lifeless*. Of the former the vinegar bacterium and yeast are good examples. Of the latter the digestive ferments, like *pepsin*, *ptyalin*, and *trypsin*, and certain vegetal ferments, like *diastase* of malt are familiar instances.

As a rule the bacteria seem to prefer neutral or slightly alkaline nitrogenous foods. They therefore decompose more readily meats, milk, and substances (such as beef-tea) made of animal matters; less readily acid fruits, timber, etc. If in the course of their activity they decompose meats, or fish, eggs, etc., with the production of evil-smelling gases or putrid odors, the process is known as *putrefaction*. Rarely, bacteria invade the animal (or plant) body and act upon the organic matters which they find there. In such cases *disease* may result, and the bacteria concerned are then known as *disease germs*.

But while bacteria appear to prefer highly organized nitrogenous (proteid) food, they are by no means dependent upon it. Experiments have shown that many species can thrive upon Pasteur's fluid, a liquid containing only ammonium tartrate and certain purely inorganic substances; and one bacterium, at least (the "nitrous"), according to Winogradsky, can thrive upon ammonium carbonate. If this proves to be true for other species, it will show that bacteria can not only obtain their nitrogen from the inorganic world, but their carbon also. Enough has been said already to prove that the bacteria are plants, for only plants can live upon inorganic food. But if the experiments just referred to are correct, bacteria are not only plants, but, in spite of their lack of chlorophyll, some at least appear to be able, like green plants to *manufacture their own food* out of the raw materials of the inorganic world. The importance of this fact in studies of the genealogy of organisms is very great, for we are no longer obliged to suppose all chlorophylless plants to be degenerate forms. They may have been the primitive forms of life.

As was the case with yeast and *Protococcus*, it is extremely difficult to make any precise statement concerning the income or outgo of bacteria. It is believed, however, that the income always includes salts and water, and the outgo CO_2, H_2O and some nitrogenous compound or, possibly, free (dissolved) nitrogen. In more favorable cases the income appears to include proteids, fats, and carbohydrates or their equivalents. Sugar is freely used under some circumstances; and fats (when saponified) and proteids peptonized, or otherwise altered, might readily be absorbed. It is probable that soluble ferments are excreted by the bacteria, which dissolve, and make absorbable, solid matters, such as meat or white of egg; and if this is true, bacteria exhibit a kind of external digestion. However this may be, it is certain that bacteria can live and multiply upon an amount of food materials so small as almost or quite to elude chemical analysis; and it is fair to say that they are among the most delicate of all reagents.

> It must not be inferred from what has been said above that bacteria are always oxidizing agents. Broadly speaking and in the long run they are such, and in this respect they resemble animals. Like the latter they are unable (because of want of chlorophyll) to utilize solar energy, and therefore must obtain their energy by oxidizing their food. Yet under certain circumstances bacteria act as reducing agents, as, for example, when they reduce nitrates to ammonia. This action only takes place, however, in the presence of organic matter, and appears to be merely an incidental effect, the oxygen of the nitrate being needed for the oxidation of carbon. What at first sight appears to be an exception, therefore, proves in the end to be a part of a general law that bacteria, like animals, are oxidizing agents, are dependent for their energy upon the potential energy of their foods, and are unable to utilize solar energy (p. 104).

It has recently been shown that many bacteria under circumstances otherwise favorable are killed by exposure to sunlight.

Related Forms. According to our present ideas of classification the bacteria form a somewhat isolated group, their nearest relatives being the slime-moulds (*Myxomycetes*) and especially the *Myxobacteria* of Thaxter, on the one hand, and the *Cyanophyceæ* the "blue-green" or "fission" algæ on the other. Neither of these, however, need be considered here.

Why Bacteria are Considered to be Plants. The bacteria were formerly regarded as infusorial animalcules (because they abound in infusions, and many have the power of active movement). They are still regarded by some as animals. Most biologists, however, regard them as plants, because they can live without proteid food (which no animal, so far as known, can do), and because in their method of reproduction and in their growth-forms they more nearly resemble the *Cyanophyceæ* than they do any animal. There is also reason to think that their cell-wall is composed of cellulose.

Bacteria and their Environment. The relations of organisms to temperature and moisture have been more thoroughly studied for the bacteria than for any other unicellular organisms on account of their bearing upon modern theories of infectious disease. In general, temperatures above 70° C. are fatal to ordinary bacteria. In general, as is shown by common experience with the "keeping" of foods in cold storage, bacteria are benumbed but not killed by moderate cold. But in special cases, particularly when they are dried slowly, bacteria may withstand even prolonged boiling or freezing or the action of poisons, so that the removal or destruction of the last traces of bacterial life is often very difficult.

Sterilization and Pasteurizing. The removal of all traces of living matter from any substance, and in particular the destruction of all bacterial life, is known as *sterilization*. To free organic substances from the larger forms of life is a comparatively easy matter; but bacteria are so minute and so ubiquitous that scarcely anything is normally free from them, and they are so hardy that it is exceedingly difficult to destroy them without at the same time destroying the substances which it is desired to sterilize. They are not normally present in the living tissues of plants or animals which are sealed against their entrance by skins or epithelia; but after these are broken or cut open (as in wounds) bacteria speedily invade the tissues. Ordinary earth, as has been said above, teems with bacteria, which are easily dried and disseminated in dust driven by the wind. Whatever is in contact, therefore, with the air or exposed to dust or dirt is never free from bacteria, and meat or milk which in the living animal are normally sterile, if exposed to the air soon become contaminated with bacteria. Sterilization (such as is required to preserve *canned goods*, for example) may be effected by heat and continued, after cooling, by exclusion of germ-laden air. *Disinfection*, which is the destruction of bacterial life by powerful poisons, is another form of sterilization. Still another is *filtration through media impervious to germs*, such as occurs in the well-known clay, or porcelain, water-filters. In the last case the pores of the filter are large enough to allow the water very slowly to pass, but too small for the bacteria.

In some cases, especially those in which disease-producing (*pathogenic*) germs may be present and yet it is impossible to use poisons and undesirable to use a high temperature, *Pasteurization* is resorted to. This consists in heating to a temperature (usually 75° C.) high enough to destroy the particular pathogenic germs supposed to be present, but not high enough to alter the digestibility or other valuable properties of the liquid in question.

For the medical, economic, and sanitary aspects of problems relating to the bacteria, reference must be had to the numerous treatises upon *Bacteriology*, perhaps the youngest, and certainly one of the most important, of the biological sciences.

CHAPTER XVII.

A HAY INFUSION.

IF a wisp of hay is put into a beaker of water and the mixture allowed to stand in a warm place there is soon formed what is known as a *hay infusion*. Microscopical examination of a drop of the liquid at the end of the first hour or two reveals little or nothing, and if the beaker be held up to the light the liquid appears clear and bright. But after some hours a marked change is found to have taken place. The liquid, originally clear, has become cloudy, and a drop of it examined microscopically will be found to be swarming with bacteria. A day or two later, the cloudiness meanwhile increasing, the microscope generally reveals not only swarms of bacteria, but also numerous infusoria. At the same time the color of the liquid has deepened, it begins to appear turbid, a scum forms on the surface, and the odor of hay, which was present at the outset, is replaced by the less agreeable odors of putrefaction. The simple experiment of bringing together hay and water has, in fact, set in motion a complicated series of physical, chemical, and biological phenomena.

The Composition of a Hay Infusion. A hay infusion consists of two principal constituents, hay and water. But neither of these is chemically pure. Hay is only dried grass which for weeks, and even months, was exposed in the field to wind and dust. Covered with the latter—often the pulverized mud of roads and roadside pools—hay is richly laden with dried bacteria and other micro-organisms; while water, such as is ordinarily drawn from a tap, frequently contains not only an abundance of free oxygen and various salts in solution, but also numerous bacteria, infusoria, algæ, diatoms, and other micro-organisms in suspension. In the making of a hay-infusion, therefore, numerous factors co-operate, and a series of complicated reactions follow one another in rapid succession. At the start both hay and water are in a state of comparative rest or equilibrium, but upon bringing them together action and reaction begin. First, the dust on the hay is wetted and soaked, and any micro-organisms in it or adhering to the hay are set free, and float in the water; next, the water finds its way into the stems and leaves of the hay, causing them to swell and resume their original form. At the same time various soluble constituents of the dead grass, such as salts, sugars, and some nitrogenous substances, diffuse outward into the water, while from such cells as have been crushed or broken open during drying or handling, solid proteid or starchy substances may pass out and mingle with the water. These simple physical reactions obviously involve a disturbance of the *chemical equilibrium* of the water. Originally able to support only a limited amount of life (such as exists in drinking-waters), it is now a soil enriched by what it has gained from the hay. The bacteria, extremely sensitive to variations in their environment, and especially to their food-supply, immediately proceed to multiply enormously, so that a biological reaction follows closely on the heels of the chemical change. But as a result of their metabolic activity the bacteria set up extensive chemical changes, which in their turn involve physical disturbances. For example, the dissolved oxygen with which the liquid was saturated soon disappears, so that more oxygen must, therefore, diffuse into the liquid from the atmosphere. Carbonic acid is generated in excess, and some may pass outwards to the air. Also, as a result of the vital activity of the micro-organisms the temperature of the infusion may rise a fraction of a degree above that of the surrounding atmosphere.

We are concerned, however, chiefly with the biological results. In consequence of the exhaustion of the oxygen supply in the lower parts of the liquid, many of the bacteria which require abundant oxygen for their growth (*aerobes*) find their way to the surface, where some pass into a kind of resting stage (*zoöglœa*) and form a scum or skin (*mycoderm*) on the surface of the liquid. Others, for which free oxygen is not necessary or to which it is even prejudicial (*anaerobes*), live and thrive in the deeper parts of the beaker. But, meantime, another phenomenon has occurred. The infusoria, originally few in number, finding the conditions favorable, have multiplied enormously, and after a day or two may be seen darting in and out among the bacteria, especially near the surface, and feeding upon them. Among the infusoria, however, are some which feed upon their fellows, so that we soon have the herbivorous infusoria pursued by carnivorous forms, the whole scene illustrating in one field of the microscope that struggle for existence which is one of the fundamental facts of biology.

Obviously, this chain of life is no stronger than its weakest part. The hay is the source of the food-supply for all these forms, and this supply must eventually become exhausted. When this happens, the bacteria cease to multiply, the herbivorous infusoria which depend upon them perish or pass into a resting stage, the carnivorous infusoria likewise starve, and all the biological phenomena must either come to an end or change their character.

Up to this point the action is purely destructive. But sooner or later microscopic green plants may appear on the scene,— *Protococcus*, it may be, or its allies,—and a constructive action begin, the waste products of the animals and of the bacteria being rebuilt by the green plants into complex organic matter. By this time, also, the dissolved organic matter will have been largely extracted from the liquid, the bacteria for the most part devoured by the infusoria, and the latter may more or less completely have given way to larger forms—to rhizopods, rotifers, small worms, and the like. The putrefying infusion has run its course, and the ordinary balance of nature has been restored.

Thenceforward an approximate equilibrium is maintained. The green plants build complex organic matter and store up the energy of light. The animals feed upon the plants, or upon one another, break down the complex matter, and dissipate energy. The ever-present bacteria break down all the refuse, extract soluble organic matter from the water, decompose the dead bodies of the animals or plants, and in the end, it may be, themselves fall victims to devouring infusoria. The physiological cycle is complete.

A hay infusion thus affords in miniature a picture of the living world. The green plants are constructive, and in the sunlight build up matters rich in potential energy. These as foods support colorless plants (such as bacteria) or animals. On these, again, herbivorous and carnivorous animals feed; and so, in the world at large, as in the hay infusion, omnivorous as well as carnivorous animals, in the long run, feed upon herbivorous animals, and the latter upon plants—either colorless or green—which thus stand as the bulwark between animals and starvation.

APPENDIX.

SUGGESTIONS FOR LABORATORY STUDIES AND DEMONSTRATIONS.

The "Laboratory Directions in General Biology," published and copyrighted by Prof. E. A. Andrews of Johns Hopkins University, will be found extremely useful and *practical*. Also the following: Huxley and Martin's "Practical Biology" (Howes and Scott), and the accompanying "Atlas of Biology," by Howes; Marshall and Hurst's "Practical Zoölogy," Colton's "Practical Zoölogy," Bumpus's "Invertebrate Zoölogy," Dodge's "Elementary Practical Biology," Brooks's "Handbook of Invertebrate Zoölogy." According to our experience, the periods for the course should be so arranged as to afford laboratory work and recitations or quizzes in about the proportions of three to two (for example, three periods of laboratory work and demonstration to two of quiz), for a half-year.

CHAPTER I. (INTRODUCTORY.)

It is convenient to give at the outset one or more practical lessons on the microscope, affording the student an opportunity to learn its different parts, use its adjustments, test the magnifying power of the various combinations, etc. A good object for a first examination is a human hair, which serves as a convenient standard of size for comparison with other things. Other good objects are starches, the scales from a butterfly's wing (sketch under different powers), a drop of milk or blood, and powdered carmine or gamboge rubbed up in water (to show the Brownian movement). The student should compare the same object as seen under the simple and the compound microscope (to show reversal of the image in the latter), and should during the course learn the use of the camera lucida (Abbe's camera, of Zeiss, the best). The stage-micrometer may also be examined at this time or later, and the student taught to prepare a scale (see Andrews) by drawing the lines, with camera, on a card under different powers (A + 2, D + 2, D + 4, of Zeiss), and labelling each with the names of lenses and actual size of the spaces, as stated on the micrometer.

Pencil-drawing should begin as soon as the first specimen is in focus, and sketches should be made, from the very first exercise onward, of everything really studied. It is absolutely indispensable *to keep a laboratory note-book*, which ought at any time to give tangible evidence that the laboratory study is bearing fruit; and in the very first laboratory exercise a beginning should be made in this direction.

The preliminary microscopy of one or two laboratory periods, corresponding to the time spent in conferences upon the first chapter of the text-book, leads naturally up to the easy microscopical studies required in connection with the second chapter.

CHAPTER II. (STRUCTURE OF LIVING ORGANISMS.)

The laboratory work may be made very brief and simple, and the facts shown largely by illustration. The principal organs of a plant and of a live or dissected animal may be shown and some of the more obvious tissues pointed out. A frog under a bell-glass, and a flowering plant (geranium) in blossom, placed side by side on the demonstration-table will serve to suggest materials for the lists of organs and the comparisons called for.

The skin of a *Calla* leaf is easily stripped off and demonstrated to the naked eye as one form of tissue. It may then be cut up and distributed for microscopic study and for proof that it is composed of cells. (During this process air is apt to replace water lost by evaporation, and must be displaced by alcohol, which in turn must be removed by water.)

For a first microscopical examination of tissue there is no better object than the leaf of a moss (a species having thin broad leaves should be chosen) or a fern prothallium. Other good objects are thin sections of a potato-tuber from *just below the surface* (stained with dilute iodine to show nuclei and starch-grains), and frog's or newt's blood, mixed with normal salt solution, and examined either fresh or slightly stained with dilute iodine.

Thin sections of pith (elder, etc.), from which the air has been displaced by alcohol, give good pictures of tissue composed of empty cells. Fresh or alcoholic muscle from the frog's leg, *gently* teased out, shows muscular tissue to be composed of elongated cells (fibres). Finally, the student may prove that he himself is composed of cells by gently scraping the inside of his lip or cheek with a scalpel, mounting the scrapings on a slide, and after adding a drop of Delafield's hæmatoxylin, covering, and examining in the usual way.

To show the lifeless matter in living tissue it suffices to examine frog's blood or human blood; sections of potatoes, especially if lightly stained with iodine; sections of geranium stems (*Pelargonium*), which usually show crystals in some of the more peripheral cells; cartilage, stained with iodine, in which the lifeless matrix remains uncolored; or prepared sections of bone, in which the spaces once filled by the living cells are now black and opaque, being filled with dust in the grinding, or with air.

CHAPTER III. (PROTOPLASM AND THE CELL.)

Naked-eye Examination of Protoplasm. A drop of protoplasm is readily obtained from one of the long (internodal) cells of *Nitella*, after removing the superfluous water and snipping off one end of the cell with scissors. The cell collapses and the drop forms at the lower (cut) end. It may be transferred to a (dry) slide and tested for its viscidity by touching it with a needle. Microscopically it is instructive chiefly by its lack of marked structure.

The Parts of the Cell. The structure of the cell is beautifully shown in properly stained and mounted preparations of unfertilized star-fish or sea-urchin eggs, or of apical buds of *Nitella*. If these are not available potato-cells or cartilage cells do very well; or sections of epithelium, glands, etc., may be shown.

The class may also mount and draw frog's or newt's blood-cells, prepared and double-stained as follows. The blood is spread out evenly on a slide and dried cautiously over a flame. Stain with hæmatoxylin for three minutes; wash thoroughly with water, add strong aqueous solution of eosin, allow to stand one minute; wash this time very rapidly, remove the excess of water *quickly* with filter-paper pressed down over the whole slide; dry rapidly, and examine with low power. If successful mount in balsam; if the specimen is not pink enough add more eosin and wash still more rapidly than before. In good specimens the cells keep their form perfectly, the cytoplasm is bright pink, and the nucleoplasm is light purple.

Epidermis from *young* leaves of hot-house lilies ("African" lily, "Chinese" lily, and especially lily-of-the-valley) yields cells showing finely the cell-wall, nucleus, and (in favorable cases) cytoplasm. If stained with acetic acid and methyl-green the nuclei are highly colored; with Delafield's hæmatoxylin the cytoplasm is more easily seen.

Cell-divisions or Cleavage are easily observed in segmenting ova or in fresh specimens of *Protococcus* (*Pleurococcus*) detached from moistened pieces of bark which bear these algæ. (See p. 178).

Stages in the cleavage of the ovum may be seen in the segmenting eggs of fresh-water snails (*Physa, Planorbis*) which are easily procured at almost any time by keeping the animals in aquaria. The old egg-masses should be removed so as to ensure the eggs being fresh. Or a supply of preserved segmenting eggs (star-fish, sea-urchin) may be kept for demonstrating the early stages.

Protoplasm in Motion. The best introduction to protoplasm in motion is afforded by a superficial examination of *Amœba* (for procuring *Amœba* see above, Chapter XII). If *Amœba* is not available young living tips of *Nitella* or *Chara* may be used. *Anacharis* and *Tradescantia* are useful, and often very beautiful, but less easy to manage, as a rule. In mounting *Nitella* or *Chara* care must be taken not to crush the cells, and as far as possible pale fresh specimens rather than darker and older ones should be chosen. If *Anacharis* is to be studied the youngest leaves should be selected from the budding ends, and not, as is sometimes recommended, leaves which are becoming yellow. The movement in the cells of *Anacharis* leaves often begins only after the leaf has been mounted for a half-hour or more; but when once established affords one of the most beautiful and striking examples of protoplasmic motion. If *Tradescantia* is to be used, care must be taken to have, if possible, flowers just open or opening. The morning is therefore preferable for work on this plant. High powers are necessary.

In all these forms the movements may often be stimulated by placing a lamp near the microscope or by *cautiously* warming the slide over the lamp-chimney. Ciliary action is easily shown in bits of the gills taken from fresh clams, mussels, or oysters, or in cells scraped from the inside of the frog's œsophagus. A striking demonstration is easily given by slitting open a frog's (or turtle's) œsophagus lengthwise, pinning out flat, moistening with normal salt solution, and placing tiny bits of moistened cork on the surface. The progressive movement of the cork-bits is then very obvious. Muscular contractility is easily shown by removing the skin from a frog's leg, dissecting out the sciatic nerve, cutting its upper end, and then stimulating the lower end, if possible, by contact with a pair of electrodes, otherwise by pinching it with forceps. If the necessary apparatus is available the regular muscle-nerve preparation may be shown (see Foster and Langley's "Practical Physiology").

Food-stuffs Contain Energy. This may be shown (in demonstrations) by sprinking *finely powdered* and *thoroughly dried* starch, sugar, or flour upon a fire, or upon a platinum dish or piece of foil heated to redness over a small flame. Oils and dried and powdered albumen (proteid) may be similarly made to burn with almost explosive violence if applied in a state of fine division in presence of air.

The Chemical Basis. (*a*) *Proteids*: *Coagulation*; *Rigor Mortis*; *Rigor Caloris*. White-of-egg may be shown (in demonstration) and made to coagulate in a test-tube hung down into a beaker of water under which is put a flame. A thermometer in the test-tube may be read off from time to time as the experiment advances, until finally coagulation begins, when the temperature is noted. The death-stiffening (*rigor mortis*) comes on very quickly in frogs killed with chloroform. Heat-stiffening (*rigor caloris*) is well shown by immersing one leg of a decapitated frog in a beaker of water at 40° C. The other leg remains normal and affords a valuable means of comparison. It is not worth while to make many chemical tests of proteids at this point.

(b) *Carbohydrates.* A useful demonstration may be made of various starches, sugars, and glycogen. The iodine-test may be applied if desired. If time allows, the microscopical appearance of potato-starch, corn-starch, Bermuda arrowroot, etc., may be dwelt upon in the laboratory-work. Cellulose is well shown in filter-paper or absorbent cotton.

(c) *Fats.* A demonstration of animal fats and vegetable oils may be made if time allows. They may be examined microscopically in a drop of milk, in an artificial emulsion made by shaking up sweet oil in dilute white-of-egg, or in fresh fatty tissue (from subcutaneous tissue of mouse, or fat-bodies of frog). It is hardly worth while to examine these substances chemically, but a few simple tests may be applied if desired.

Dialysis. A demonstration of dialysis is easily made by inverting a broken test-tube, tying the membrane over the flaring end, filling the tube to a marked point with strong salt or glucose solution, and immersing it in a beaker of distilled water. After an hour or so the fluid will be found to have risen in the test-tube against gravity.

Temperature and Protoplasm. The profound influence of temperature on protoplasm is well shown by the frog's heart. Decapitate a frog and destroy the spinal cord. Expose the heart and count the beats at the room temperature. Then pour upon the heart iced normal salt solution. Again count the beats. Next pour upon it normal salt solution heated to $35°$ C. The number of beats will follow the fall and rise of temperature.

CHAPTERS IV TO VIII. (THE EARTHWORM.)

Large earthworms must be used or satisfactory results cannot be expected. Pains should therefore be taken to procure the large *L. terrestris* (*not* the common *Allolobophora mucosa*), which is readily recognizable by the flattened posterior end. This species is not everywhere common; hence a supply should be procured and kept in a cool place in barrels half full of earth, on the surface of which is placed a quantity of moss. They will thus live for months. *L. terrestris* may be obtained in great numbers between April and November, by searching for them at night with a lantern in localities where numerous castings show them to abound (a rather heavy but rich soil will be found most productive). They will then be found extended from their burrows, lying on the surface of the ground, and may be seized with the fingers. Considerable dexterity is needed, and it is necessary to tread very softly or the worms take alarm and instantly withdraw into their burrows.

For dissection fresh specimens are far preferable for most purposes, though *properly* preserved ones answer the purpose. Fresh specimens should be nearly killed by being placed for a short time (about five minutes) in 70% alcohol, and then stretched out to their utmost extent in 50% alcohol in a dissecting-pan, the two ends being fastened by pins. They should then be at once cut open along the middle dorsal line with scissors, the flaps pinned out, and the dissection continued under the 50% alcohol. (They must be *completely* covered with the liquid.) By this method the minutest details of structure may be observed, and many of the dissections should be done under a watchmaker's lens.

For preservation (every detail of which should be attended to) a number of living worms are placed in a broad vessel filled to a depth of about an inch with water. A little alcohol is then cautiously dropped on the surface of the water at intervals until the worms are stupefied and become perfectly motionless and relaxed (this may require an hour or two). They are then transferred to a large shallow vessel containing just enough 50% alcohol to cover them, and are carefully straightened out and arranged side by side. After an hour the weak alcohol is replaced by stronger (70%), which should be changed once or twice at intervals of a few hours; they are finally placed in 90% alcohol, which should be *liberally used*. The trouble demanded by this method will be fully repaid by the results. The worms should be quite straight, fully extended, and plump, and they may be used either for dissection or for microscopic study.

For the purposes of section-cutting worms should be carefully washed and placed in a moist vessel containing plenty of wet filter-paper torn into shreds. The worms will devour the paper, which should be changed several times, until the paper is voided perfectly clean. The worms are then preserved in the ordinary way, and when properly hardened are cut into short pieces, stained with borax-carmine, imbedded in paraffin, and cut into sections with the microtome.

The living worms should first be observed—their shape, movements, behavior to stimuli, pulsation of the dorsal vessel (time the pulse and vary the rate by temperature changes). Well-preserved specimens should then be carefully studied for the external characters (draw through the fingers to feel the setæ). (Sketch.) Observe openings. The nephridial openings cannot be seen, but if preserved worms be soaked some hours in water and the cuticle peeled off they may be clearly seen in this. A general dissection of a fresh specimen should now be made, and the positions of the larger organs studied. (Make partial sketch, to be filled out afterwards, as in Fig. 24.) The alimentary canal and circulatory organs should now be carefully studied. Even the smallest of the blood-vessels may easily be worked out under the lens by using fresh specimens (killed in 70% alcohol and afterwards dissected under *water*) and carefully turning aside the alimentary canal.

The alimentary canal should afterwards be cut through behind the gizzard and gradually dissected away in front, exposing the nerve-cord and the reproductive organs (wash away dirt with a pipette). No great difficulty should be found in making out any of the parts, excepting the testes. These are difficult to find in mature worms, but may be found with ease in those which have no median seminal vesicles (usually the case with specimens having no clitellum).

The contents of the seminal receptacles and vesicles from a fresh worm should be examined with the microscope. Remove an ovary (with forceps and small curved scissors), mount in water, and study. (Stained in alum-carmine and mounted in balsam the ovary is a beautiful object.) The student should also remove a fresh nephridial funnel and part of a nephridium, and study with the microscope. (This may have to be shown by the demonstrator, but should never be omitted, as the ciliary action is one of the most striking things to see.) A careful dissection of the anterior part of the nervous system should also be made.

If time presses, the detailed study of microscopical sections may be omitted, but a series of prepared sections should be kept on hand and a demonstration given.

The embryological development is too difficult to study, but very instructive demonstrations may be given by those who have had some experience. In the neighborhood of Philadelphia egg-capsules may be found in great numbers in old manure-heaps, in May and June. One end of the capsule should be sliced off with a very sharp scalpel and the contents drawn out, under water, with a large-mouthed pipette. The mass may then be mounted in water under a supported cover-glass and studied with the microscope. The embryos may be preserved in Perenyi's fluid, and either studied whole in the preserving fluid or hardened in alcohol and cut into series of sections.

CHAPTERS IX TO XI. (THE COMMON BRAKE.)

Except when the ground is frozen *Pteris* may be dug up and brought into the laboratory in a fresh state. Fronds may be cut and dried in midsummer and considerably freshened (by a moment's immersion in warm water) when needed to be used (in the opening exercise) to illustrate the aerial portion of the plant. Rhizomes may be obtained at convenience and kept in weak alcohol (50%).

The Morphology of the Body. To illustrate this, one *whole and entire* plant should, if possible, be at hand for examination. The aerial and the underground portions may then be sketched in their normal relations. Branches, roots, and old leaf-stalks should be pointed out, identified, and sketched.

The Anatomy of the Rhizome should first be made out with the naked eye. The lateral ridges will be detected by the class, which should be asked to draw the cross-section as seen with the naked eye. For this preliminary work each student should have a piece of rhizome two or three inches in length. (Care should afterwards be taken that the drawing has been correctly placed dorsoventrally.) A rough dissection with jack-knife or large scalpel may next follow, with inferences as to the characters of the several tissues found (as fibrous, pulpy, woody, etc.).

The Microscopic Anatomy of the Rhizome is interesting, and, for the most part, easy, but demands much time. If time allows, cross-sections of roots may be made and mounted in balsam. They are readily cut in pith. Sections of the rhizome may be made freehand with a razor or, better, with a microtome: but the old stems are exceedingly hard and liable to injure the knives.

The Frond or Leaf may be obtained in fruit in July and August and preserved in alcohol. From it sections of leaflets may easily be got by imbedding in pith. Epidermis is obtained with some difficulty (by beginners) after scraping. Fresh fern-leaves from hothouses answer the purpose as well, are easier to get, and more attractive. Really good sections of fern-leaves are not easy for beginners to make. They should be kept on hand.

Sporangia may be obtained in abundance from alcoholic specimens of *Pteris*, or upon hothouse ferns, even in midwinter. Some of the many species of *Pteris* found in hot-houses answer every purpose. The thin edge of a scalpel slipped under the unripe indusium removes the latter, and generally also long ranks of sporangia in all stages of development. In some sporangia spores may be found. Sporangia and spores are always readily got, but care must be taken to select fruit-dots which are not too old or too young.

Sprouting the Spores. To obtain good specimens of sprouting spores and prothallia *free from dirt*, we can recommend the following procedure: Fill several small flower-pots, which have been thoroughly cleaned inside and out, with clean fine sand. Sterilize the whole by baking in an oven or a hot-air sterilizer. Set the pots into large (porcelain) dishes capable of holding water, and keep the bottom of these dishes covered to the depth of one inch with water; cover the pots completely with bell-glasses. After twenty-four hours, or after the sand and the pots have become thoroughly wet, inside *and outside*, dust thickly the sand *and the outsides* of the pots with spores (obtained from fern-houses by shaking fertile fronds over white paper). Care must be taken to get *spores*, and not merely empty sporangia. After a week or longer (sometimes several weeks) a bit of the surface-layer of sand is removed to a drop of water on a slide and examined for sprouting spores. These will often be found in various stages of development. After a month or two prothallia will appear on the *outside* of the pots; and as these are clean, they may be removed and examined (bottom side upwards) free of all dirt.

Failing these, prothallia may almost always be found in fern-houses on the tops or sides of the pots, and especially on the moist earth under the benches. Care should be taken not to confound prothallia with the lighter green and relatively coarse liverwort (*Lunularia*) often found in hothouses.

The Sexual Organs of Prothallia. With good clean specimens these are easily found with a rather low power. Higher powers are needed to make out details. If the archegonia and and antheridia are young they are green; if old, brown. On young prothallia antheridia only are often found, and on very old ones archegonia only.

Fertilization. This is not easy to observe, but the attempt may be made by examining successively a number of very fresh and vigorous prothallia in different stages. They must be mounted carefully (not flooded with water), and spermatozoids are generally more easily found swimming about after the specimen has been mounted a little while.

Embryology. Except in its general features, this is too difficult for the beginner. He may, however, observe the later stages by studying old prothallia with the young fern just appearing, and young ferns with the old prothallia still adherent.

Chlorophyll and Starch. Vigorous prothallia afford excellent examples of cells bearing chlorophyll-bodies in which *starch* is easily detected. Some of the marginal cells should be examined with the highest power, attention being given to the chlorophyll-bodies and their arrangement. In favorable cases one may observe the opaque rod-like or oval grains inside the latter, and prove by reagents that they are starch grains.

The student should also examine, at this point, the large chromatophores of *Nitella*, which may be obtained by pressing out a drop of the contents from an internodal cell, adding *dilute* iodine solution, and examining with a high power. In favorable cases as many as a dozen starch grains, stained blue, may be found inside a single elliptical chlorophyll-body.

Chapter XII. (Amœba.)

Amœba is one of the most capricious of animals, appearing and disappearing with inexplicable suddenness, and as a rule it cannot be found at the time when needed, unless special preparations have been made in advance. It is never safe to trust to chance for a supply of material. It is equally unsafe to trust to the methods usually prescribed. Amœbæ may, however, often be procured in abundance and with tolerable certainty as follows: A month or six weeks beforehand collect considerable quantities of water-plants (especially *Nitella* or *Chara*) from various pools or slow ditches, with an abundance of sediment from the bottom. It is important to select clear, quiet pools containing an abundance of organic matter (such as desmids, diatoms, etc., in the sediment)—not temporary rain-pools or such as are choked with inorganic mud (dirt washed in by rain). The material thus procured should be distributed in numerous (10 to 20) open shallow dishes (earthenware milk-pans) and allowed to stand about the laboratory in various places—some exposed to the sun, others in the shade. The contents of many, perhaps all, of the vessels will undergo putrefactive changes and swarm with life—first with bacteria, later with infusoria—and will then gradually become clear again as in a hay-infusion. The sediment should now be examined at intervals, and *Amœba* are almost certain to appear, sooner or later, in one or more of the vessels. Usually the small *A. radiosa* appears first, but these should only be used if it is found impossible to procure *A. Proteus*, which is far larger, clearer, and more interesting. Experience will show that particular pools always yield a crop of *Amœba*, while others do not. When once a productive source is found all trouble is ended.

If possible a sediment should be selected that swarms with *Amœba*. It is very discouraging for students to pass most of their time looking *for* the animals instead of *at* them. Large cover-glasses should be used, and the material taken with a pipette from the very surface of the sediment (not from its deeper layers). When first mounted the animals are usually contracted, and only become fully extended after a time. Outline sketches should be made at stated intervals, the structure of the protoplasm carefully studied, the pulse of the contractile vacuole timed (vary by varying temperature), and the effect of tapping the cover-glass noted. It is practically useless to look for fission, for encysted forms, or for the external opening of the contractile vacuole; and the ingulfing of food or passing out of waste matters is rarely seen. The formation of pseudopodia should be carefully studied. After examining the living animals they should be killed and stained with dilute iodine.

Arcella is almost always, and *Difflugia* sometimes, found with *Amœba*. These forms may be examined for comparison.

It is desirable also to compare white blood-corpuscles, which may be obtained either by pricking the finger or, better, from a frog or newt. A drop of blood, received upon a slightly warmed slide, should be covered and sealed with oil around the edge of the cover-glass. The white corpuscles are at first rounded, but soon begin to show change of form. (No contractile vacuole, no differentiation into ectoplasm and entoplasm, often no nucleus visible.)

CHAPTER XII. (INFUSORIA.)

Paramœcia are almost certain to appear in the earlier stages of the *Amœba* cultures, and in similar decomposing liquids or infusions, and to ensure having them a large number of vessels and jars containing an excess of vegetable matter should be prepared a month or more beforehand. Their successful study is very easy if they are procured *in very large numbers* (the water should be milky with them), otherwise it is practically impossible. Three slides of them should be prepared and set aside for a short time (under cover, preferably, in a moist chamber) to allow the animals to become quiet. One slide should contain simply a drop of the infusorial water; a second the same, with the addition of a little powdered carmine; to the third add a drop or two of an aqueous solution of chloral hydrate (made by dropping a crystal or two into a watch-glass of water). The first slide should be studied first; and it will usually be found that after a time the animals crowd about the edges of the cover, often lying nearly or quite still. If this is not the case, the specimens paralyzed by chloral may be studied. The carmine specimens will show beautiful food-vacuoles filled with carmine; and by careful study the formation of the vacuoles may be observed.

The general structure should be carefully studied, the contractile vacuoles particularly examined (they are seen best in dying specimens or in those paralyzed by chloral), and dividing or conjugating individuals looked for (they are often abundant). The only really difficult point is the nucleus, which cannot be well seen in the living animal. It may be clearly seen by mounting a drop, to which a little dilute iodine or 2% acetic acid has been added. The former shows the cilia well, the latter the trichocysts. Osmic acid and corrosive sublimate also give good preservation. The internal changes during fission and conjugation must be studied in prepared specimens mounted in balsam. Such preparations are often of great beauty and interest.

Vorticella must be sought for on duck-weed or other plants, or on floating sticks, and the like. *Zoöthamnion*, *Carchesium*, etc., are liable to appear at any time in the aquaria. All these forms are easily studied. Conjugation is very rarely seen, but fission and motile forms are common. The macronucleus is especially well shown in dead or dying specimens.

Chapter XIV. (Protococcus.)

Protococcus (*Pleurococcus*) is found in abundance on the northerly side of old trees in many parts of the United States. In case it cannot be obtained in any region it may be procured, during 1895 and 1896, from Prof. Sedgwick, Institute of Technology, Boston, Mass., by mail. The laboratory-work with it is too easy to require comment. See, however, Arthur, Barnes & Coulter's "Plant Dissection" (Henry Holt & Co., New York).

Chapter XV. (Yeast.)

Bakers', brewers', compressed, and dried yeast may be had in the markets. Brewers' yeast is to be preferred, as compressed yeast-cakes contain starch, bacteria, and other extraneous matters. All of the kinds may be cultivated to good advantage in wort (to be obtained at breweries) or in Pasteur's fluids. (See Huxley and Martin, chapter on Yeast.) Wild yeasts may be found by examining sweet cider microscopically. For the following methods of demonstrating nuclei in yeast and obtaining ascospores we are indebted to Mr. S. C. Keith, Jr.

To Demonstrate Nuclei in Yeast. Any good actively-growing yeast will answer, but a large (brewers') yeast is preferable. Mix a little of the yeast with an equal amount of tap-water in a test-tube and shake thoroughly. Add an equal volume of Hermann's fluid and shake again. As soon as the yeast has settled pour off the supernatant liquid and wash the yeast by decantation. Transfer some of the cells to a slide, fix by drying, stain by Heidenhain's iron-haematoxylin method (see *Centralblatt für Bacteriologie*, xiv. (1893), pp. 358–360), wash, dehydrate with alcohol, follow with cedar-oil, and mount in balsam. In successful specimens the effect is very satisfactory. (See Fig. 96.)

A Simpler Method. To demonstrate nuclei in yeast more quickly and very easily the following method may be used: Boil (in a test-tube) for a moment an infusion of *very vigorous* yeast in water, place a drop of the boiled infusion on a slide, add a drop of *very dilute* "Dahlia" solution, cover, and after one or two minutes examine with a high power. The nuclei in most of the cells will be easily discoverable.

To Obtain Ascospores in Yeast. It has been usually recommended to employ for this purpose blocks of plaster-of-Paris. We have found the following method more trustworthy:

The yeast to be used should be the "top" yeast used in ale-breweries. It should also be actively growing and fresh. If fresh yeast cannot be obtained, some may be revived by cultivation for 24 hours at 25° C. in wort, and a little of the thick sedimentary portion may then be placed in a very thin layer on dry filter-paper which has previously been sterilized by baking. The filter-paper is then placed on a layer of cotton about $\frac{1}{4}$ inch in thickness lying on a plate or saucer, the cotton having previously been thoroughly wetted with cold sterilized tap-water. The whole is covered by a bell-glass and set in a rather warm place (25° C.). In the course of two or three days spores will be found in many of the cells. The lower the temperature the longer is the time required for spore formation. If "bottom" yeast is used instead of "top" yeast a much longer time is required, and the results are far more uncertain.

Chapter XVI. (Bacteria.)

For the study of Bacteria it is very desirable to have a large species, and for this purpose there is none better than *Bacillus megaterium*, which may be obtained from almost any bacteriological laboratory and grown in the bouillon used by bacteriologists. During 1895 and 1896 it may be obtained from Boston (see above). This form is very large, and produces spores readily. (See De Bary, "Lectures on Bacteria;" Sternberg, "Bacteriology;" Abbott, "Principles of Bacteriology;" etc.) The prolonged study of bacteria is not suited to beginners. *Vinegar bacteria* may be seen in the mother-of-vinegar by pressing a bit of it out under a cover slip and examining with a high power. The jelly of mother-of-vinegar is a good example of *zoöglœa*. The white scum which appears on aquaria and infusions is of the same general character (*zoöglœa*).

Chapter XVII. (A Hay Infusion.)

To make a successful hay infusion care should be taken to use water containing numerous and various organisms, and therefore distilled water, spring-waters, and well-waters, are in general to be avoided. Tap-water should also be avoided if it is derived from springs or wells. The best water for the purpose is that drawn from ponds, rivers, lakes, or other *surface* sources. Clean ditch or pool water is excellent. The choice of hay is less important, but it is well to avoid old hay and hay that is very woody. The infusion should be warmed, but not heated or boiled. It may be kept in a beaker in diffuse daylight, e.g., in a north window, the beaker being loosely covered.

INSTRUMENTS AND UTENSILS.*

The student should have access to the following articles:

A compound microscope with two eyepieces and low and high power objectives (i.e., about 1 in. and $\frac{1}{5}$ in., or objectives

---

* Most of the apparatus and reagents here mentioned may be obtained from any first-class dealer in physical and microscopical apparatus, e.g., from The

A and D of Zeiss, or ½ and ⅛ inch of Bausch and Lomb; still higher powers are desirable).

A simple dissecting microscope; a desirable form is an ordinary watchmaker's lens provided with a support. An ordinary pocket-lens; glass slides (3 × 1 in.), cover-glasses, watch-crystals, small gummed labels, needles with adjustable handles, camel's-hair brushes, blotting and filter paper, a good razor, pipettes (medicine-droppers), glass rods and tubes, glass or porcelain dishes for staining, etc., a set of small dissecting instruments (small scalpel, forceps, and straight-pointed scissors), a section-lifter, pieces of pith for section-cutting, thread, a shallow tin pan lined with wax, long insect pins for pinning out dissected specimens, drawing materials, and a note-book for sketches and other records.

Each table should be furnished with a set of small reagent-bottles, a Bunsen burner, wash-bottle, test-tubes, beakers, and a bell-glass for protection from dust. Thermometers, a balance, microtome, drying oven, and a paraffin water-bath should also be accessible.

REAGENTS AND TECHNICAL METHODS.*

Alcohol.—Since biological laboratories belonging to incorporated institutions obtain alcohol duty free, it should be *liberally supplied* and freely used. Alcohol of 100°, i.e., "absolute" alcohol, may be purchased in 1-pound bottles. "Squibb's" absolute alcohol may be obtained of any druggist,† but ordinary alcohol of 90–95% answers nearly every purpose. "Cologne spirits," i.e., alcohol of about 94%, may be obtained from the distillers at 60c., or thereabouts, per gallon. It may then be

---

Bausch & Lomb Optical Co., Rochester, N. Y.; the Franklin Educational Co., Hamilton Place, Boston; or Queen & Co., Chestnut Street, Philadelphia. Chemical and other apparatus may be obtained from Eimer & Amend, 205–211 Third Avenue, N. Y.

* Every laboratory should be supplied with some of the standard books upon this subject, e.g., Strasburger's *Botanische Practicum*, Jena; Whitman's *Methods of Research in Microscopical Anatomy and Embryology*, Boston; Lee, *The Microtomist's Vade Mecum*, last edition; Zimmerman's *Botanical Microtechnique* (Humphrey), Holt, N. Y.

† See also Whitman, l. c., p. 14.

diluted to 80%, 70%, 50%, etc., as needed. For this purpose an alcoholimeter is very convenient.

Acetic Acid.—One or two parts glacial acetic acid to 100 parts water.

Acetic Acid and Methyl-green.—This is valuable for staining nuclei in vegetal tissues. Dissolve methyl-green in one or two per cent acetic acid until a rich deep color is obtained.

Borax-carmine.—Add to a 4% aqueous solution of borax 2-3% carmine, and heat until the carmine dissolves. Add an equal volume of 70% alcohol, and filter after 24 hours. After staining (6-12 hours, or more for large objects, a few minutes for sections) place the object in acidulated alcohol (100 c.c. 35% alcohol, 3-4 drops hydrochloric acid) and leave until the color turns from dull to bright red (10-30 m.). Afterwards remove to 70% alcohol.

Canada Balsam, Mounting in.—This invaluable substance may be obtained in the crude condition, dried by prolonged heating, and then dissolved in chloroform, benzole, or turpentine, for use. The benzole solution is perhaps the best, and may be obtained from most of the dealers. The principles of mounting in balsam are very simple. It does not mix with water or alcohol, but mixes freely with clove-oil, chloroform, benzole, etc. Objects are therefore generally treated, first with very strong alcohol, 95-100%, in order to remove the water; then with clove-oil, chloroform, or turpentine to remove the alcohol, and afterwards mounted in a drop of balsam. This should usually be placed on the cover-glass, which is thereupon inverted over the object. The balsam gradually sets and the preparations are permanently preserved.

Carmine.—Carmine may be obtained as a powder, which when rubbed up thoroughly with water in a mortar passes into a state of very fine subdivision. This property makes it available for experiments with cilia, etc.

It is more often used in solution, as a staining agent. (See **Borax-carmine.**)

Cellulose-test.—Saturate the object in iodine solution, wash in water, and place it in strong sulphuric acid prepared by carefully pouring 2 volumes of the concentrated acid into 1 volume of water.

Collodion and Clove-oil.—Used for fixing sections to the slide in order to prevent the displacement of delicate or isolated parts in balsam-mounting. Mix one part of ether-collodion and three parts of oil of cloves. In mounting, varnish a slide with the mixture by means of a camel's-hair brush, lay on the sections, and place the slide for a few minutes on the water-bath (i.e., until the clove-oil evaporates). Transfer the slide to a wide-mouthed bottle of turpentine (to dissolve the paraffin), remove it and drain off the turpentine, place a drop of Canada balsam on the middle of a cover-glass, and invert it over the object.

Dahlia.—Dissolve in water.

Eosin.—Dissolve in water until a bright-red solution is obtained. It should be diluted when used.

Glycerine, dilute.—Two parts glycerine, one part distilled water.

Hæmatoxylin (Delafield's).—Add 4 c.c. of saturated alcoholic solution of hæmatoxylin to 150 c.c. of strong aqueous solution of ammonia-alum; let the mixture stand a week or more in the light, filter, and add 25 c.c. of glycerine and 25 c.c. of methyl alcohol. The fluid improves greatly after standing some weeks or months.

Hæmatoxylin (Kleinenberg's).—To a saturated solution of calcium chloride in 70% alcohol add an excess of *pure* alum; filter after 24 hours and add 8 volumes of 70% alcohol, filtering again if necessary. Add a saturated alcoholic solution of hæmatoxylin until the liquid becomes purple-blue. The longer the liquid stands before using, the better. It should be diluted for use with the alum-calcium-chloride solution in 70% alcohol.

Hermann's Fluid.—See Lee's *Vade Mecum*.

Iodine Solution.—Dissolve potassium iodide in a small quantity of water, add metallic iodine until the mixture assumes a dark-brown color, and then dilute to a dark-sherry color. The solution should be kept from the light.

Magenta (Aniline Red).—Dissolve in water.

Methyl Green.—Used in aqueous or alcoholic solution or with acetic acid.

Normal Fluid (Normal Salt Solution).—Dissolve 7.50 grams of sodium chloride in 1 litre of distilled water.

Paraffin.—"Hard" and "soft" paraffins, i.e., those of high and low melting-points, should be mixed in such proportions that the melting-point lies between 50° and 55° C.

Perenyi's Fluid.—Ten-per-cent nitric acid 4 parts, 90% alcohol 3 parts, ½% aqueous solution of chromic acid 3 parts. Not to be used until the mixture assumes a violet hue. Leave objects in the fluid 30 minutes to an hour, then 24 hours in 70% alcohol, and finally place in 90 per cent alcohol.

Schultze's Macerating Fluid.—Dissolve a gram of potassium chlorate in 50 c.c. of nitric acid. The tissue should be boiled in the mixture and afterwards thoroughly washed in water.

Schulze's Solution.—Dissolve zinc in pure hydrochloric acid, evaporate in the presence of metallic zinc, on a water-bath, to a syrupy consistency, add as much iodide of potassium as will dissolve, and then saturate with iodine. (When heated with this fluid cellulose turns blue.

Section-cutting.—Many objects can be cut by hand with a razor (which must be very sharp). The object should be held in the left hand while the razor is pointed away from the body, and allowed to rest on the tips of the fingers with its edge turned towards the left. It is then drawn gently towards the body so as gradually to shave off the section. Small objects may be held between two pieces of watchmaker's pith previously soaked in water. In either case the razor should be kept wet.

Many objects, however, require more careful treatment by one of the following methods:

A. *Paraffin Method.*—After hardening and staining, the object is soaked in strong alcohol (95% or more) until the water is thoroughly extracted (2-12 hours, changing the alcohol at least once), then in chloroform until the alcohol is extracted (2-12) hours), and then in melted paraffin (not warmer than 55° C.) on a water-bath for 15 to 30 minutes (too high a temperature or too long a bath causes excessive shrinkage). Some of the paraffin is then poured into a small paper-box, or into adjustable metal frames. The object is transferred to it and after the mass has begun to set it is placed in cold water until quite hard. It is then cemented (by paraffin) to a square piece of cork and placed in the section-cutter or microtome.

The sections may be cut singly with the oblique knife or by

the ribbon-method,* the knife being kept dry in either case. In mounting they should be fixed by the collodion-method. (See **Collodion** and **Clove-oil**.)

B. *Celloidin Method.*—This is especially applicable to delicate vegetal tissues. After dehydrating the object thoroughly in alcohol, soak it 24 hours in a mixture of equal parts of alcohol and ether. Make a thick solution of celloidin in the same mixture and soak the object for some hours in it. It may then be imbedded as follows: Dip the smaller end of a tapering cork in the celloidin solution, allow it to dry for a moment (blowing on it if necessary), and then build upon it a mass of celloidin, allowing it to dry a moment after each addition. Transfer the object to the cork and cover it thoroughly with the celloidin. Then float the cork in 82–85% (0.842 sp. gr.) alcohol until the mass has a firm consistency (24 h.). It may then be cut in the microtome with the oblique knife, which must be kept dripping with 82–85% alcohol. Keep the sections in 82–85% alcohol until ready to mount them, then soak them for a minute in strong alcohol, transfer to a slide, pour on chloroform until the alcohol is removed, drain off the liquid, quickly add a drop of balsam, and cover. (See also Whitman, l. c., p. 113.)

---

* See Whitman, l. c. p. 71.

INDEX.

Absorption, 48, 52, 101, 165.
Accretion, 166.
Achromatin, 23.
Actinophrys, 166.
Adaptation, 97, 98, 144.
Adventitious buds, 130.
Ærobes, 202.
Ætiology, 6.
Agamogenesis, 73, 130, 163.
Albuminous bodies, 36.
Alimentation, 48, 105.
Alimentary canal, 82, 92.
Alimentary system, 49.
Allolobophora, 41.
Alternation of generations, 130.
Amœba, 27, 158, 216.
Amœboid cells, 64.
Amphiaster, 84.
Amphimixis, 168.
Anabolism, 33, 100, 149, 164.
Anacharis, 29.
Anærobes, 202.
Anatomy, 7.
Animalcule, 158, 199.
Annulus, 132.
Anus, 46, 82, 165.
Antheridia, 135.
Aortic arches, 54, 55.
Apical buds, 111, 116, 123.
Apical cell, 123.
Apogamy, 143.
Apospory, 143.
Arcella, 166.
Archegonia, 137.
Archenteron, 80, 82, 85.
Archesporium, 131.
Archoplasm, 79, 80.
Arthrospore, 195.
Ascospore, 187.
Asexual reproduction, 73.
Assimilation, 182.
Aster, 79, 84.
Attraction sphere, 83, 84.
ATWATER, W. O., 34.

Bacilli, 192.
Bacteria, 64, 178, 192.
Bast-fibres, 120.

Biology, 1, 6, 7, 8.
Bisexual, 73, 130.
Blastopore, 80, 85.
Blastosphere, 85.
Blastula, 80, 90.
Blood, 15, 16, 90, 102.
Blood-vessels, 54.
Blue-green algæ, 183, 192.
Body, 19, 24, 84, 107, 156.
Body-cavity, 47.
Bone, 16.
Botany, 6, 7.
Branches, 111, 122, 130.
Branchiæ, 62.
Budding, 186.
Bursaria, 176.

Calciferous glands, 51.
CALKINS, G. N., 171.
Capillaries, 54.
Capsules of eggs, 78.
Capsulogenous glands, 46.
Carbohydrates, 37, 101.
Carchesium, 176.
Carnivora, 177, 203.
Cartilage, 15, 16.
Castings, 42, 53.
Cell, 12, 20.
Cell-division, 24, 83.
Cell-theory, 20.
Cellulose, 37.
Cell-wall, 22, 23.
Centrosome, 79, 83, 84.
Cerebral ganglia, 65, 69.
Chalk, 166.
Chara, 24.
Chemiotaxis, 139.
Chlorococcus, 178.
Chloragogue-cells, 52, 61, 93.
Chlorophyll, 126, 151, 215.
Chlorophyll-bodies, 179, 215.
Chroöcoccus, 183.
Chromatin, 23, 83.
Chromatophores, 147, 179.
Chromosomes, 83, 84.
Cilia, 31, 63, 74, 137, 192.
Circulation, 48, 53, 101, 165.
CLAPARÈDE, 96.

INDEX

Classification, 7.
Clitellum, 46, 77, 78, 88, 92.
Coagulation, 36, 39.
Cocci, 192.
Cœlenterata, 88.
Cœlom, 47, 82.
Cœlomic fluid, 53.
Cohn, 21.
Cold storage, 199.
Colloidal, 36.
Colony, 176.
Commissures, 65.
Conjugation, 171, 181.
Connective tissue, 70, 90.
Consciousness, 69, 70.
Contractility, 62, 164.
Coördination, 48, 64, 67, 164.
Copulation, 77.
Cross-fertilization, 74.
Crystals, 17.
Cushion, 135.
Cuticle, 71, 91.
Cyanophyceæ, 183, 192, 199.
Cyclical change, 5, 72, 89.
Cytoplasm, 22, 84.

Darwin, 42, 51, 70, 99, 103.
Death, 152.
De Bary, 115, 143.
Defæcation, 53, 165.
Desmids, 178, 183.
Dialysis, 36, 210.
Diastatic ferment, 52.
Diatoms, 178, 183.
Dichogamy, 138.
Differentiation, 11, 84, 141.
Differentiation, antero-posterior, 43, 110.
Differentiation, dorso-ventral, 43, 110.
Differentiation of the tissues, 25.
Difflugia, 166.
Digestion, 48, 49, 52, 101, 165.
Diplococcus, 194.
Disease-germs, 192, 197.
Disinfection, 200.
Dissepiments, 47, 94.
Distribution, 7.
Division of labor, 11, 26, 156, 165.
Dorsal pore, 48.
Dorsal vessel, 54.
Dujardin, 21.

Earthworm, 41.
Ectoblast, 81.
Ectoplasm, 158.
Egg, 24.
Egg laying, 77.
Egg-nucleus, 79.
Egg-string, 74.
Embryo, 25.
Embryology, 7, 72, 78.

Endospore, 187, 194.
Endosporium, 134.
Energy, 32, 99, 146, 151.
Entoblast, 81.
Entoplasm, 158.
Environment, 97, 103, 144, 151.
Epidermal system, 114.
Epidermis, 114, 116.
Epistylis, 176.
Epithelium, 90.
Euglena, 176.
Excretion, 48, 53, 59, 100, 165.
Exosporium, 134.
Eye-spot, 176.

Fæces, 53.
Farlow, 143.
Fats, 17, 37, 101.
Feathers, 18.
Ferns, 105.
Ferment, 52.
Fermentation, 191, 197.
Fertilization, 73, 78, 139.
Fibro-vascular system, 114.
Fibro-vascular bundles, 142.
Filtration, 200.
Fission, 163.
Flagellum, 176, 192.
Fol, 79.
Foods, 146.
Foraminifera, 166.
Fore-gut, 86.
Foster, Michael, 153, 163.
Frédericq, 52.
Frond, 125.
Functions, 9.
Fundamental system, 114.
Fungi, 147.

Gamete, 181.
Gamogenesis, 73, 130, 168.
Ganglion, 64, 94.
Gastrula, 80.
Gastrulation, 84.
Germ-cells, 24, 73, 90, 130.
Germination, 134.
Germ-layers, 81, 84, 85.
Germ-layer theory, 88.
Germ-plasm, 89, 152.
Germinal spot, 74.
Germinal vesicle, 74.
Giant-fibres, 94.
Gills, 62.
Girdle, 78.
Gizzard, 51, 71.
Glæocapsa, 178, 183.
Glucose, 52.
Glycogen, 37.
Gregarina, 64.
Growth, 165.
Guard-cells, 128.

INDEX.

Hæmatococcus, 178.
Hæmoglobin, 54.
Hair, 18.
Hay infusion, 201.
Herbivora, 176, 203.
Heredity, 84.
Hermaphrodite, 73, 130.
HERTWIG, 79.
Hibernation, 38.
Hind-gut, 86.
Histology, 7.
HOOKE, ROBERT, 20.
HOOKER, SIR W. J., 106
HOPPE-SEYLER, 35.
HUXLEY, 2, 4.
Hypodermis, 92.

Impregnation, 73, 139.
Individual, 13, 156, 164.
Indusium, 131.
Infusions, 168.
Infusoria, 168, 217.
Inheritance, 80, 84.
Intussusception, 4, 165.
Irritability, 164.

JOHNSON, 35.

Katabolism, 33, 99, 149, 164.
Karyokinesis, 83.
KEITH, S. C., Jr., 186, 195.
KRUKENBERG, 52.

Lateral ridges, 111, 114.
Leaf, 11, 125.
LENHOSSÉK, 95.
Leptothrix, 194.
LINNÆUS, 105.
Lumbricus, 41.
Lungs, 62.
Lymph, 53.
Lymph-cells, 64.

Macrogamete, 175.
Macro nucleus, 170, 171.
Malic acid, 139.
MAUPAS, 170.
Meristem, 123.
Mesoblast, 81.
Mesophyll, 126.
Metabolism, 33, 100, 101, 148, 164.
Metamerism, 45.
METCHNIKOFF, 53.
Microgamete, 175.
Micronucleus, 170, 171.
Micro-organisms, 201.
Middle-piece, 74, 79, 80.
Mid-gut, 86.
Mitosis, 83.
MOHL, H. VON, 21.
Morphology, 6, 7.

Mother-of-vinegar, 194, 195.
Mother-cells, 134, 137.
Motion, 48.
Motor system, 62.
Mouth, 49, 80, 85, 165.
Muscles, 14, 26, 27, 62, 90.
MULDER, 35.
Mycoderma, 194, 202.
Myxobacteria, 199.
Myxomycetes, 199.

Natural selection, 99.
Nephridia, 58, 59.
Nerves, 64, 90.
Nerve-cells, 94.
Nerve-centre, 68.
Nerve-impulses, 67.
Nervous system, 64, 82, 94, 102.
Nitella, 28.
Nitrogen, 147.
Nucleolus, 23.
Nucleoplasm, 22.
Nucleus, 16, 23, 186.
Nutrition, 99, 146.

Œsophagus, 18.
Old age, 72, 152, 166.
Oöphore, 130.
Oösphere, 73, 138.
Oöspore, 139.
Organisms, 9.
Organogeny, 85.
Organs, 9.
Ovaries, 74.
Oviduct, 75.
Ovum, 73, 74, 89.

Paramœcium, 168.
Parasites, 192.
Parenchyma, 116.
PASTEUR, 188.
Pasteurization, 200.
Pasteur's fluid, 189, 197.
Pathogenic, 200.
Pathology, 6, 7.
Peptic ferment, 52.
Peptone, 52, 101.
Peristaltic actions, 51, 54, 55.
PFEFFER, 139.
Phagocytes, 53, 61, 64, 158.
Pharyngeal ganglia, 67.
Pharynx, 49.
Physiological properties of protoplasm, 163, 182, 183.
Physiology, 6, 7, 166.
Physiology of the nervous system, 67.
Polar cells, 79.
Pole-cells, 82.
Poisons, 39.
Plasma, 53.
Pleurococcus, 178.

Primordial utricle, 29.
Proctodæum, 82, 86.
Pronucleus, 79.
Prosenchyma, 116.
Prostomium, 45.
Protection, 71.
Proteids, 3, 33, 52.
Proteus animalcule, 27, 158.
Prothallium, 130, 135, 214.
Protococcus, 178.
Protonema, 134.
Protoplasm, 16, 20, 207, 208.
Protozoa, 158.
Pseudopodia, 27, 158·
Psychology, 7, 8.
Pulse, 54.
Putrefaction, 197, 201.
PURKINJE, 21.

Radiolaria, 166.
Receptacle, 131.
Receptaculum ovorum, 75.
Reflex action, 67.
Regeneration, 73.
Reproduction, 48, 72, 111, 130, 152, 165.
Respiration, 61, 150, 165.
RETZIUS, 95.
Rhizoids, 134.
Rhizome, 111, 140.
Rhizopoda, 166.
Rigor caloris, 39.
Rigor mortis, 209.
Roots, 122.

Saccharomyces, 184.
SACHS, 115.
Salivary glands, 51.
Sap, 14.
Saprophytes, 192.
Sarcina, 194.
Schizomycetes, 192.
SCHLEIDEN, 20.
SCHULTZE, MAX, 21.
SCHWANN, 20.
Sciences, biological, 1, 6.
Sciences, physical, 1.
Segmentation, 24, 80.
Segmentation cavity, 84, 85.
Seminal receptacle, 77.
Seminal vesicle, 76.
Sensation, 48.
Sense organs, 42, 69.
Senses, 42, 69.
Sensitive system, 69.
Setæ, 46, 63.
Setigerous glands, 63, 77.
Sexual reproduction, 73.
Sieve-tubes, 116.
Sight, 42, 69, 70
Skin, 128.
Slipper animalcule, 168.

Smell, 42, 69.
Sociology, 7, 8.
Somatic cells, 73.
Somatic layer, 85.
Somatopleure, 82, 86.
Somites, 45.
SPENCER, HERBERT, 3, 99, 146.
Spermaries, 74, 75.
Spermatosphere, 77.
Spermatozoid, 137.
Spermatozoön, 73, 74
Sperm-duct, 76.
Sperm-nucleus, 79.
Spiderwort, 29.
Spirilla, 192.
Splanchnopleure, 82, 86.
Spontaneous generation, 33.
Sporangia, 130.
Spores, 24, 130, 194.
Sporophore, 130.
Staphylococcus, 194.
Starch, 17, 37, 146.
Stentor, 170.
Sterilization, 199.
Stimulus, 67.
Stipe, 125.
Stomach-intestine, 51.
Stomata, 126, 128.
Stomodæum, 82, 86.
Streptococcus, 194.
Struggle for existence, 203.
Stylonichia, 170.
Sugar, 37.
Sun-animalcule, 166.
Survival of the fittest, 99.
Symbiosis, 177.
Symmetry, bilateral, 44, 110.
Symmetry, serial, 45.
Sympathetic system, 67.

Taste, 42, 69, 70.
Taxonomy, 7.
Temperature, 38, 199, 210.
Testes, 74, 75.
Tissues, 11, 13.
Touch, 42, 69, 70.
Toxicology, 39.
Tracheæ, 116.
Tracheids, 116.
Tradescantia, 29.
Transpiration, 146.
Trichocysts, 168.
Tryptic ferment, 52.
Twins, 88.
Typhlosole, 51, 91.

Unicellular animals, 158.
Unicellular organisms, 156, 177.
Unicellular plants, 178.

Vacuoles, 24, 162, 170.

Vascular system, 54.
Vas deferens, 76.
Veins, 126.
VEJDOVSKY, 79, 81
Venation, 129.
Vessels, 116.
Vinegar, 196.
VIRCHOW, 21.
Vital energy, 33.
Vital force, 33.
Vitellus, 74, 78.
Vorticella, 168, 172.

WHITE, 42.

White blood-cells, 64.
Whirlpool, 2.
WINOGRADSKY, 197.

Yeast, 178.
Yeast, bottom, 190.
Yeast, red, 191.
Yeast, top, 190.
Yeast, wild, 190.

Zoöglœa, 194, 195.
Zoöids, 176.
Zoölogy, 6, 7.
Zoöspores, 181.
Zoöthamnion, 176.

www.ingramcontent.com/pod-product-compliance
Lightning Source LLC
Chambersburg PA
CBHW031741230426
43669CB00007B/427